Reprint Series No. 2

Environment, Heredity, and Intelligence

Compiled from the
Harvard
Educational
Review

Seventh Printing, 1975
Library of Congress Catalog Card Number 71-87869
Printed in the United States of America
Designed and printed by Capital City Press, Montpelier, Vermont 05602
Harvard Educational Review
Longfellow Hall, 13 Appian Way
Cambridge, Massachusetts 02138

Environment, Heredity, and Intelligence

How Much Can We Boost IQ and Scholastic Achievement?

ARTHUR R. JENSEN

University of California, Berkeley

Arthur Jensen argues that the failure of recent compensatory education efforts to produce lasting effects on children's IQ and achievement suggests that the premises on which these efforts have been based should be reexamined.

He begins by questioning a central notion upon which these and other educational programs have recently been based: that IQ differences are almost entirely a result of environmental differences and the cultural bias of IQ tests. After tracing the history of IQ tests, Jensen carefully defines the concept of IQ, pointing out that it appears as a common factor in all tests that have been devised thus far to tap higher mental processes.

Having defined the concept of intelligence and related it to other forms of mental ability, Jensen employs an analysis of variance model to explain how IQ can be separated into genetic and environmental components. He then discusses the concept of "heritability," a statistical tool for assessing the degree to which individual differences in a trait like intelligence can be accounted for by genetic factors. He analyzes several lines of evidence which suggest that the heritability of intelligence is quite high (i.e., genetic factors are much more important than environmental factors in producing IQ differences).

After arguing that environmental factors are not nearly as important in determining IQ as are genetic factors, Jensen proceeds to analyze the environmental influences which may be most critical in determining IQ. He concludes

Harvard Educational Review Vol. 39 No. 1 Winter 1969, 1–123 *that prenatal influences may well contribute the largest environmental influence on IQ. He then discusses evidence which suggests that social class and racial varia-tions in intelligence cannot be accounted for by differences in environment but must be attributed partially to genetic differences.*

After he has discussed the influence of the distribution of IQ in a society on its functioning, Jensen examines in detail the results of educational programs for young children, and finds that the changes in IQ produced by these programs are generally small. A basic conclusion of Jensen's discussion of the influence of environment on IQ is that environment acts as a "threshold variable." Extreme environmental deprivation can keep the child from performing up to his genetic potential, but an enriched educational program cannot push the child above that potential.

Finally, Jensen examines other mental abilities that might be capitalized on in an educational program, discussing recent findings on diverse patterns of mental abilities between ethnic groups and his own studies of associative learning abilities that are independent of social class. He concludes that educa-tional attempts to boost IQ have been misdirected and that the educational process should focus on teaching much more specific skills. He argues that this will be accomplished most effectively if educational methods are developed which are based on other mental abilities besides I.Q.

Because of the controversial nature of Dr. Jensen's article, the Spring Issue of the Review will feature a discussion of the article by five psychologists: Carl Bereiter, Lee Cronbach, James Crow, David Elkind, and J. McVicker Hunt. Readers are also invited to react.

The Failure of Compensatory Education

Compensatory education has been tried and it apparently has failed.

Compensatory education has been practiced on a massive scale for several years in many cities across the nation. It began with auspicious enthusiasm and high hopes of educators. It had unprecedented support from Federal funds. It had theoretical sanction from social scientists espousing the major underpinning of its rationale: the "deprivation hypothesis," according to which academic lag is mainly the result of social, economic, and educational deprivation and dis-crimination—an hypothesis that has met with wide, uncritical acceptance in the atmosphere of society's growing concern about the plight of minority groups and the economically disadvantaged.

The chief goal of compensatory education—to remedy the educational lag of disadvantaged children and thereby narrow the achievement gap between "minority" and "majority" pupils—has been utterly unrealized in any of the large compensatory education programs that have been evaluated so far. On the basis of a nationwide survey and evaluation of compensatory education programs, the United States Commission on Civil Rights (1967) came to the following conclusion:

The Commission's analysis does not suggest that compensatory education is incapable of remedying the effects of poverty on the academic achievement of individual children. There is little question that school programs involving expenditures for cultural enrichment, better teaching, and other needed educational services can be helpful to disadvantaged children. The fact remains, however, that none of the programs appear to have raised significantly the achievement of participating pupils, as a group, within the period evaluated by the Commission. (p. 138)

The Commission's review gave special attention to compensatory education in majority-Negro schools whose programs "were among the most prominent and included some that have served as models for others." The Commission states: "A principal objective of each was to raise the academic achievement of disadvantaged children. Judged by this standard the programs did not show evidence of much success" (p. 138).[1]

Why has there been such uniform failure of compensatory programs wherever they have been tried? What has gone wrong? In other fields, when bridges do not stand, when aircraft do not fly, when machines do not work, when treatments do not cure, despite all conscientious efforts on the part of many persons to make them do so, one begins to question the basic assumptions, principles, theories, and hypotheses that guide one's efforts. Is it time to follow suit in education?

---

[1] Some of the largest and most highly publicized programs of compensatory education that have been held up as models but which produced absolutely no significant improvement in the scholastic achievement of disadvantaged students are: the *Banneker Project* in St. Louis (8 years), *Higher Horizons* in New York (5 years), More Effective Schools in New York (3 years), and large-scale programs in Syracuse, Seattle, Philadelphia, Berkeley, and a score of other cities (for detailed reports see U.S. Commission on Civil Rights, 1967, pp. 115-140).

Reports on Project Head Start indicate that initial gains of 5 to 10 points in IQ on conventional intelligence tests are a common finding, but this gain usually does not hold up through the first year of regular schooling. More positive claims for the efficacy of Head Start involve evidence of the detection and correction of medical disabilities in disadvantaged preschool children and the reportedly favorable effects of the program on children's self-confidence, motivation, and attitudes toward school.

The theory that has guided most of these compensatory education programs, sometimes explicitly, sometimes implicitly, has two main complementary facets: one might be called the "average children concept," the other the "social deprivation hypothesis."

The "average children" concept is essentially the belief that all children, except for a rare few born with severe neurological defects, are basically very much alike in their mental development and capabilities, and that their apparent differences in these characteristics as manifested in school are due to rather superficial differences in children's upbringing at home, their preschool and out-of-school experiences, motivations and interests, and the educational influences of their family background. All children are viewed as basically more or less homogeneous, but are seen to differ in school performance because when they are out of school they learn or fail to learn certain things that may either help them or hinder them in their school work. If all children could be treated more alike early enough, long before they come to school, then they could all learn from the teacher's instruction at about the same pace and would all achieve at much the same level, presumably at the "average" or above on the usual grade norms.

The "social deprivation hypothesis" is the allied belief that those children of ethnic minorities and the economically poor who achieve "below average" in school do so mainly because they begin school lacking certain crucial experiences which are prerequisites for school learning—perceptual, attentional, and verbal skills, as well as the self-confidence, self-direction, and teacher-oriented attitudes conducive to achievement in the classroom. And they lack the parental help and encouragement needed to promote academic achievement throughout their schooling. The chief aim of preschool and compensatory programs, therefore, is to make up for these environmental lacks as quickly and intensively as possible by providing the assumedly appropriate experiences, cultural enrichment, and training in basic skills of the kind presumably possessed by middle-class "majority" children of the same age.

The success of the effort is usually assessed in one or both of two ways: by gains in IQ and in scholastic achievement. The common emphasis on gains in IQ is probably attributable to the fact that it can be more efficiently "measured" than scholastic achievement, especially if there is no specific "achievement" to begin with. The IQ test can be used at the very beginning of Headstart, kindergarten, or first grade as a "pre-test" against which to assess "post-test" gains. IQ gains, if they occur at all, usually occur rapidly, while achievement is a long-term affair. And probably most important, the IQ is commonly interpreted as indicative of a more general kind of intellectual ability than is reflected by the acquisition of specific scholastic knowledge and skills. Since the IQ is known to predict scholastic performance better than any other single measurable attribute of the child, it is believed, whether rightly or wrongly, that if the child's IQ can be appreciably raised, academic achievement by and large will take care of itself, given normal motivation and standard instruction. Children with average or above-average IQs generally do well in school without much special attention. So the remedy deemed logical for children who would do poorly in school is to boost their IQs up to where they can perform like the majority—in short to make them all at least "average children." Stated so bluntly, the remedy may sound rather grim, but this is in fact essentially what we are attempting in our special programs of preschool enrichment and compensatory education. This simple theme, with only slight embellishments, can be found repeated over and over again in the vast recent literature on the psychology and education of children called culturally disadvantaged.

So here is where our diagnosis should begin—with the concept of the IQ: how it came to be what it is; what it "really" is; what makes it vary from one individual to another; what can change it, and by what amount.

The Nature of Intelligence

The nature of intelligence is one of the vast topics in psychology. It would be quite impossible to attempt to review here the main theoretical issues and currents of thought in this field. Large volumes have been written on the subject (e.g., Guilford, 1967; Stoddard, 1943), to say nothing of the countless articles. An enlightening brief account of the history of the concept of intelligence has been presented by Sir Cyril Burt (1968). The term "intelligence," as used by psychologists, is itself of fairly recent origin. Having been introduced as a technical term in psychology near the turn of the century, it has since filtered down into common parlance, and therefore some restriction and clarification of the term as it will be used in the following discussion is called for.

Disagreements and arguments can perhaps be forestalled if we take an operational stance. First of all, this means that probably the most important fact about intelligence is that we can measure it. Intelligence, like electricity, is easier to measure than to define. And if the measurements bear some systematic relationships to other data, it means we can make meaningful statements about the phenomenon we are measuring. There is no point in arguing the question to which there is no answer, the question of what intelligence *really* is. The best we can do is to obtain measurements of certain kinds of behavior and look at their relationships to other phenomena and see if these relationships make any kind of sense and order. It is from these orderly relationships that we can gain some understanding of the phenomena.

But how did the instruments by which we measure intelligence come about in the first place? The first really useful test of intelligence and the progenitor of nearly all present-day intelligence tests was the Metrical Scale of Intelligence devised in 1905 by Binet and Simon. A fact of great but often unrealized implications is that the Binet-Simon test was commissioned by the Minister of Public Instruction in Paris for the explicit purpose of identifying children who were likely to fail in school. It was decided they should be placed in special schools or classes before losing too much ground or receiving too much discouragement. To the credit of Binet and Simon, the test served this purpose quite well, and it is now regarded as one of the major "breakthroughs" in the history of psychology. Numerous earlier attempts to devise intelligence tests were much less successful from a practical standpoint, mainly because the kinds of functions tested were decided upon in terms of early theoretical notions about the basic elements of "mind" and the "brass instrument" laboratory techniques for measuring these elemental functions of consciousness, which were then thought to consist of the capacity for making fine sensory discriminations in the various sensory modalities. Although these measurements were sufficiently reliable, they bore little relationship to any "real life" or "common sense" criteria of behavior ranging along a "dull"— "bright" continuum. The psychological sagacity of Binet and Simon as test constructors derived largely from their intimate knowledge and observation of the behavior of young children and of what, precisely, teachers expected of them in school. Binet and Simon noted the characteristics distinguishing those children described by their teachers as "bright" from those described as "dull," and, from these observations and considerable trial-and-error, they were finally able to make up a graded series of test items that not only agreed with teachers' judgments of children's scholastic capabilities but could make the discriminations more finely and more accurately than any single teacher could do without prolonged observation of the child in class. The Binet-Simon scale has since undergone many revisions and improvements, and today, in the form developed by Terman, known as the Stanford-Binet Intelligence Scale, it is generally regarded as the standard for the measurement of intelligence.

But the important point I wish to emphasize here is that these Binet tests, and in effect all their descendants, had their origin in the educational setting of the Paris schools of 1900, and the various modifications and refinements they have undergone since then have been implicitly shaped by the educational traditions of Europe and North America. The content and methods of instruction represented in this tradition, it should be remembered, are a rather narrow and select sample of all the various forms of human learning and of the ways of imparting knowledge and skills. The instructional methods of the traditional classroom were not invented all in one stroke, but evolved within an upper-class segment of the European population, and thus were naturally shaped by the capacities, culture, and needs of those children whom the schools were primarily intended to serve. At least implicit in the system as it originally developed was the expectation that not all children would succeed. These methods of schooling have remained essentially unchanged for many generations. We have accepted traditional instruction so completely that it is extremely difficult even to imagine, much less to put into practice, any radically different forms that the education of children could take. Our thinking almost always takes as granted such features as beginning formal instruction at the same age for all children (universally between ages five and six), instruction of children in groups, keeping the same groups together in lock step fashion through the first several years of schooling, and an active-passive, showing-seeing, telling-listening relationship between teacher and pupils. Satisfactory learning occurs under these conditions only when children come to school with certain prerequisite abilities and skills: an attention span long enough to encompass the teacher's utterances and demonstrations, the ability voluntarily to focus one's attention where it is called for, the ability to comprehend verbal utterances and to grasp relationships between things and their symbolic representations, the ability to inhibit large-muscle activity and engage in covert "mental" activity, to repeat instruction to oneself, to persist in a task until a self-determined standard is attained—in short, the ability to engage in what might be called self-instructional activities, without which group instruction alone remains ineffectual.

The interesting fact is that, despite all the criticisms that can easily be leveled at the educational system, the traditional forms of instruction have actually worked quite well for the majority of children. And the tests that were specifically devised to distinguish those children least apt to succeed in this system have also proved to do their job quite well. The Stanford-Binet and similar intelligence tests predict various measures of scholastic achievement with an average validity coefficient of about .5 to .6, and in longitudinal data comprising intelligence test and achievement measures on the same children over a number of years, the multiple correlation between intelligence and scholastic achievement is almost as high as the reliability of the measures will permit.

The Generality and Limitations of Intelligence

If the content and instructional techniques of education had been markedly different from what they were in the beginning and, for the most part, continue to be, it is very likely that the instruments we call intelligence tests would also have assumed a quite different character. They might have developed in such a way as to measure a quite different constellation of abilities, and our conception of the nature of intelligence, assuming we still called it by that name, would be correspondingly different. This is why I think it so important to draw attention to the origins of intelligence testing.

But in granting that the measurement and operational definitions of intelligence had their origins in a school setting and were intended primarily for scholastic purposes, one should not assume that intelligence tests measure *only* school learning or cultural advantages making for scholastic success and fail to tap anything of fundamental psychological importance. The notion is sometimes expressed that psychologists have mis-aimed with their intelligence tests. Although the tests may predict scholastic performance, it is said, they do not *really* measure intelligence—as if somehow the "real thing" has eluded measurement and perhaps always will. But this is a misconception. We *can* measure intelligence. As the late Professor Edwin G. Boring pointed out, intelligence, by definition, is what intelligence tests measure. The trouble comes only when we attribute more to "intelligence" and to our measurements of it than do the psychologists who use the concept in its proper sense.

The idea of intelligence has justifiably grown considerably beyond its scholastic connotations. Techniques of measurement not at all resembling the tasks of the Binet scale and in no way devised with the idea of predicting scholastic performance can also measure approximately the same intelligence as measured by the Binet scale. The English psychologist Spearman devoted most of his distinguished career to studying the important finding that almost any and every test involving any kind of complex mental activity correlates positively and substantially with any and every other test involving complex mental activity, regardless of the specific content or sensory modality of the test. Spearman noted that if the tests called for the operation of "higher mental processes," as opposed to sheer sensory acuity, reflex behavior, or the execution of established habits, they showed positive intercorrelations, although the tests bore no superficial resemblance to one another. They might consist of abstract figures involving various spatial relationships, or numerical problems, or vocabulary, or verbal analogies. For example, a vocabulary test shows correlations in the range of .50 to .60 with a test that consists of copying sets of designs with colored blocks; and a test of general information correlates about .50 with a test that involves wending through a printed maze with a pencil. Countless examples of such positive correlations between seemingly quite different tests can be found in the literature on psychological tests. Spearman made them the main object of his study. To account for the intercorrelations of "mental" tests, he hypothesized the existence of a single factor common to all tests involving complex mental processes. All such tests measure this common factor to some degree, which accounts for the intercorrelations among all the tests. Spearman called the common factor "general intelligence" or simply g. And he invented the method known as factor analysis to determine the amount of g in any particular test. He and his students later developed tests, like Raven's Progressive Matrices and Cattell's Culture Fair Tests of g, which measure g in nearly pure form. We should not reify g as an entity, of course, since it is only a hypothetical construct intended to explain covariation among tests. It is a hypothetical source of variance (individual differences) in test scores. It can be regarded as the nuclear operational definition of intelligence, and when the term intelligence is used it should refer to g, the factor common to all tests of complex problem solving.

In examining those tests most heavily loaded with g, Spearman characterized the mental processes which they seemed to involve as "the ability to educe relations and correlates"—that is, to be able to see the general from the particular and the particular as an instance of the general. A similar definition of intelligence was expressed by Aquinas, as "the ability to combine and separate"—to see the difference between things which seem similar and to see the similarities between things which seem different. These are essentially the processes of abstraction and conceptualization. Tasks which call for problem solving requiring these processes are usually the best measures of g. Despite numerous theoretical attacks on Spearman's basic notion of a general factor, g has stood like a rock of Gibralter in psychometrics, defying any attempt to construct a test of complex problem solving which excludes it.

Standard intelligence scales such as the Binet and the Wechsler are composed of a dozen or so subtests which differ obviously in their superficial appearance: vocabulary, general information, memory span for digits, block designs, figure copying, mazes, form boards, and so on. When the intercorrelations among a dozen or more such tests are subjected to a factor analysis or principal components analysis, some 50 percent or more of the total individual differences variance in all the tests is usually found to be attributable to a general factor common to all the tests. Thus, when we speak of intelligence it is this general factor, rather than any single test, that we should keep in mind.

Attempts to assess age differences in intelligence or mental development which rely on complex techniques that bear little formal resemblance to the usual intelligence tests still manage to measure g more than anything else. Piaget's techniques for studying mental growth, for example, are based largely on the child's development of the concepts of invariance and conservation of certain properties —number, area, and volume. When a large variety of Piaget tasks are factor analyzed along with standard psychometric tests, including the Stanford-Binet and Raven's Progressive Matrices, is it found that the Piaget tasks are loaded on the general factor to about the same extent as the psychometric tests (Vernon, 1965). That is to say, children fall into much the same rank order of ability on all these cognitive tests. Tuddenham (1968) has developed a psychometric scale of intelligence based entirely upon Piaget's theory of cognitive development. The test makes use of ten of the techniques developed by Piaget for studying conservation, seriation, reversal of perspective, and so on. Performance on these tasks shows about the same relationship to social class and race differences as is generally found with the Stanford-Binet and Wechsler scales. It seems evident that what we call general intelligence can be manifested in many different forms and thus permits measurement by a wide variety of techniques. The common feature of all such intercorrelated tests seems to be their requirement of some form of "reasoning" on the part of the subject—some active, but usually covert, transformation or manipulation of the "input" (the problem) in order to arrive at the "output" (the answer).

The conceptually most pure and simple instance of this key aspect of intelligence is displayed in the phenomenon known as cross-modal transfer. This occurs when a person to whom some particular stimulus is exposed in one sensory modality can then recognize the same stimulus (or its essential features) in a different sensory modality. For example, show a person a number of differently shaped wooden blocks, then point to one, blindfold the person, shuffle the blocks, and let the person find the indicated block by using his sense of touch. Or "write" in bold strokes any letter of the alphabet between a child's shoulder blades. It will be a completely unique stimulus input for the child, never encountered before and

10

never directly conditioned to any verbal response. Yet, most children, provided they already know the alphabet, will be able to name the letter. There are no direct neural connections between the visual and the tactile impressions of the stimulus, and, although the child's naming of the letter has been conditioned to the visual stimulus, the tactile stimulus has been associated with neither the visual stimulus nor the verbal response. How does the child manage to show the cross modal transfer? Some central symbolic or "cognitive" processing mechanism is involved, which can abstract and compare properties of "new" experiences with "old" experiences and thereby invest the "new" with meaning and relevance. Intelligence is essentially characterized by this process.

Is g Unitary or Divisible?

It is only when the concept of g is attributed meaning above and beyond that derived from the factor analytic procedures from which it gains its strict technical meaning that we run into the needless argument over whether g is a unitary ability or a conglomerate of many subabilities, each of which could be measured independently. We should think of g as a "source" of individual differences in scores which is common to a number of different tests. As the tests change, the nature of g will also change, and a test which is loaded, say, .50 on g when factor analyzed among one set of tests may have a loading of .20 or .80, or some other value, when factor analyzed among other sets of tests. Also, a test which, in one factor analysis, measures only g and nothing else, may show that it measures g and one or more other factors when factor analyzed in connection with a new set of tests. In other words, g gains its meaning from the tests which have it in common. Furthermore, no matter how simple or "unitary" a test may appear to be, it is almost always possible to further fractionate the individual differences variance into smaller subfactors. I have been doing this in my laboratory with respect to a very simple and seemingly "unitary" ability, namely, digit span (Jensen, 1967b). Changing the rate of digit presentation changes the rank order of subjects in their ability to recall the digits. So, too, does interposing a 10-second delay between presentation and recall, and interpolating various distractions ("retroactive inhibition") between presentation and recall, and many other procedural variations of the digit span paradigm. Many—but, significantly, not all—of these kinds of manipulations introduce new dimensions or factors of individual differences. It is likely that when we finally get down to the irreducible "atoms" of memory span ability, so to speak, if we ever do get there, the elements that make up memory span ability will not themselves even resemble what we think of as abilities in the usual sense of the term. And so probably the same would be true not only for digit span, but for any of the subtests or items that make up intelligence tests.

A simple analogy in the physical realm may help to make this clear. If we are interested in measuring general athletic ability, we can devise a test consisting of running, ball throwing, batting, jumping, weight lifting, and so on. We can obtain a "score" on each one of these and the total for any individual is his "general athletic ability" score. This score would correspond to the general intelligence score yielded by tests like the Stanford-Binet and the Wechsler scales.

Or we can go a step further in the refinement of our test procedure and intercorrelate the scores on all these physical tasks, factor analyze the intercorrelations, and examine the general factor, if indeed there is one. Assuming there is, we would call it "general athletic ability." It would mean that on all of the tasks, persons who excelled on one also tended to be superior on the others. And we would note that some tasks were more "loaded" with this general factor than others. We could then weight the subtest scores in proportion to their loading on g and then add them up. The total, in effect, is a "factor score," and gives us a somewhat more justifiable measure of "general athletic ability," since it represents the one source of variation that all the athletic skills in our test battery share in common.

To go still further, let us imagine that the running test has the highest loading on g in this analysis. To make the issue clear-cut, let us say that all its variance is attributable to the g factor. Does this mean that running ability is not further analyzable into other components? *No, it simply means that the components into which running can be analyzed are not separately or independently manifested in either the running test or the other tests in the battery.* But we can measure these components of running ability independently, if we wish to: total leg length, the ratio of upper to lower leg length, strength of leg muscles, physical endurance, "wind" or vital capacity, ratio of body height to weight, degree of mesomorphic body build, specific skills such as starting speed—all are positively correlated with running speed. And if we intercorrelate these measures and factor analyze the correlations, we would probably find a substantial general factor common to all these physical attributes, name it what you will. We could combine the measures on these various physical traits into a weighted composite score which would predict running ability as measured by the time the person takes to cross the finish line. The situation seems very similar to the analysis of the psychological processes that make up "general intelligence."

Fluid and Crystallized Intelligence

Raymond B. Cattell (1963) has made a conceptually valid distinction between two aspects of intelligence, *fluid* and *crystallized*. Standard intelligence tests generally measure both the fluid and crystallized components of *g*, and, since the two are usually highly correlated in a population whose members to a large extent share a common background of experience, culture, and education, the fluid and crystallized components may not always be clearly discernible as distinct factors. Conceptually, however, the distinction is useful and can be supported empirically under certain conditions. *Fluid* intelligence is the capacity for new conceptual learning and problem solving, a general "brightness" and adaptability, relatively independent of education and experience, which can be invested in the particular opportunities for learning encountered by the individual in accord with his motivations and interests. Tests that measure mostly fluid intelligence are those that minimize cultural and scholastic content. Cattell's Culture Fair Tests and Raven's Progressive Matrices are good examples. *Crystallized* intelligence, in contrast, is a precipitate out of experience, consisting of acquired knowledge and developed intellectual skills. Fluid and crystallized intelligence are naturally correlated in a population sharing a common culture, because the acquisition of knowledge and skills in the first place depends upon fluid intelligence. While fluid intelligence attains its maximum level in the late teens and may even begin to decline gradually shortly thereafter, crystallized intelligence continues to increase gradually with the individual's learning and experience all the way up to old age.

Occupational Correlates of Intelligence

Intelligence, as we are using the term, has relevance considerably beyond the scholastic setting. This is so partly because there is an intimate relationship between a society's occupational structure and its educational system. Whether we like it or not, the educational system is one of society's most powerful mechanisms for sorting out children to assume different roles in the occupational hierarchy.

The evidence for a hierarchy of occupational prestige and desirability is unambiguous. Let us consider three sets of numbers.[2] First, the Barr scale of occupations, devised in the early 1920s, provides one set of data. Lists of 120 representative occupations, each definitely and concretely described, were given to 30 psychological judges who were asked to rate the occupations on a scale from 0 to 100

[2] I am indebted to Professor Otis Dudley Duncan (1968, pp. 80-100) for providing this information.

according to the grade of intelligence each occupation was believed to require for ordinary success. Second, in 1964, the National Opinion Research Center (NORC), by taking a large public opinion poll, obtained ratings of the *prestige* of a great number of occupations; these prestige ratings represent the average standing of each occupation relative to all the others in the eyes of the general public. Third, a rating of socioeconomic status (SES) is provided by the *1960 Census of Population: Classified Index of Occupations and Industries,* which assigns to each of the hundreds of listed occupations a score ranging from 0 to 96 as a composite index of the average income and educational level prevailing in the occupation.

The interesting point is the set of correlations among these three independently derived occupational ratings.

The Barr scale and the NORC ratings are correlated .91.

The Barr scale and the SES index are correlated .81.

The NORC ratings and the SES index are correlated .90.

In other words, psychologists' concept of the "intelligence demands" of an occupation (Barr scale) is very much like the general public's concept of the prestige or "social standing" of an occupation (NORC ratings), and both are closely related to an independent measure of the educational and economic status of the persons pursuing an occupation (SES index). As O. D. Duncan (1968, pp. 90-91) concludes, "... 'intelligence' is a socially defined quality and this social definition is not essentially different from that of achievement or status in the occupational sphere.... When psychologists came to propose operational counterparts to the notion of intelligence, or to devise measures thereof, they wittingly or unwittingly looked for indicators of capability to function in the system of key roles in the society." Duncan goes on to note, "Our argument tends to imply that a correlation between IQ and occupational achievement was more or less built into IQ tests, by virtue of the psychologists' implicit acceptance of the social standards of the general populace. Had the first IQ tests been devised in a hunting culture, 'general intelligence' might well have turned out to involve visual acuity and running speed, rather than vocabulary and symbol manipulation. As it was, the concept of intelligence arose in a society where high status accrued to occupations involving the latter in large measure, so that what we now *mean* by intelligence is something like the probability of acceptable performance (given the opportunity) in occupations varying in social status."

So we see that the prestige hierarchy of occupations is a reliable objective reality in our society. To this should be added the fact that there is undoubtedly some relationship between the levels of the hierarchy and the occupations' intrinsic interest, desirability, or gratification to the individuals engaged in them. Even if all occupations paid alike and received equal respect and acclaim, some occupations would still be viewed as more desirable than others, which would make for competition, selection, and, again, a kind of prestige hierarchy. Most persons would agree that painting pictures is more satisfying than painting barns, and conducting a symphony orchestra is more exciting than directing traffic. We have to face it: the assortment of persons into occupational roles simply is not "fair" in any absolute sense. The best we can ever hope for is that true merit, given equality of opportunity, act as the basis for the natural assorting process.

Correlation Between Intelligence and Occupational Achievement.

Because intelligence is only one of a number of qualities making for merit in any given occupation, and since most occupations will tolerate a considerable range of abilities and criteria of passable performance, it would be surprising to find a very high correlation between occupational level and IQ. Although the rank order of the *mean* IQs of occupational groups is about as highly correlated with the occupations' standing on the three "prestige" ratings mentioned above as the ratings are correlated among themselves, there is a considerable dispersion of IQs *within* occupations. The IQ spread increases as one moves down the scale from more to less skilled occupations (Tyler, 1965, pp. 338-339). Thus, the correlation, for example, between scores on the Army General Classification Test, a kind of general intelligence test, and status ratings of the civilian occupations of 18,782 white enlisted men in World War II was only .42. Since these were mostly young men, many of whom had not yet completed their education or established their career lines, the correlation of .42 is lower than one would expect in the civilian population. Data obtained by the U.S. Employment Service in a civilian population shows a correlation of .55 between intelligence and occupational status, a value which, not surprisingly, is close to the average correlation between intelligence and scholastic achievement (Duncan, et al., 1968, pp. 98-101). Although these figures are based on the largest samples reported in the literature and are therefore probably the most reliable statistics, they are not as high as the correlations found in some other studies. Two studies found, for example, that IQs of school boys correlated .57 and .71 with their occupational status 14 and 19 years later, respectively (Tyler, 1965, p. 343). It is noteworthy that the longer interval showed the higher correlation.

Duncan's (1968) detailed analysis of the nature of the relationship between intelligence and occupational status led him to the conclusion that "the bulk of the influence of intelligence on occupation is indirect, via education." If the correlation of intelligence with education and of education with occupation is, in effect, "partialled out," the remaining "direct" correlation between intelligence and occupation is almost negligible. But Duncan points out that this same type of analysis (technically known as "path coefficients analysis") also reveals the interesting and significant finding that intelligence plays a relatively important part as a cause of differential *earnings*. Duncan concludes: ". . . men with the same schooling and in the same line of work are differentially rewarded in terms of mental ability" (1968, p. 118).

Correlations Between Intelligence and Job Performance Within Occupations

Intelligence, via education, has its greatest effect in the assorting of individuals into occupational roles. Once they are in those roles, the importance of intelligence per se is less marked. Ghiselli (1955) found that intelligence tests correlate on the average in the range of .20 to .25 with ratings of actual proficiency on the job. The speed and ease of training for various occupational skills, however, show correlations with intelligence averaging about .50, which is four to five times the predictive power that the same tests have in relation to work proficiency *after* training. This means that, once the training hurdle has been surmounted, many factors besides intelligence are largely involved in success on the job. This is an important fact to keep in mind at later points in this article.

Is Intelligence "Fixed"?

Since the publication of J. McV. Hunt's well-known and influential book. *Intelligence and Experience* (1961), the notion of "fixed intelligence" has assumed the status of a popular cliché among many speakers and writers on intelligence, mental retardation, cultural disadvantage, and the like, who state, often with an evident sense of virtue and relief, that modern psychology has overthrown the "belief in fixed intelligence." This particular bugaboo seems to have loomed up largely in the imaginations of those who find such great satisfaction in the idea that "fixed intelligence" has been demolished once and for all.

Actually, there has been nothing much to demolish. When we look behind the rather misleading term "fixed intelligence," what we find are principally two real and separate issues, each calling for empirical study rather than moral philosophizing. Both issues lend themselves to empirical investigation and have long been subjects of intensive study. The first issue concerns the genetic basis of individual differences in intelligence; the second concerns the stability or constancy of the IQ throughout the individual's lifetime.

Genotype and Phenotype. Geneticists have avoided confusion and polemics about the issue of whether or not a given trait is "fixed" by asking the right question in the first place: how much of the variation (i.e., individual differences) in a particular trait or characteristic that we observe or measure (i.e., the *phenotype*) in a given population can we account for in terms of variation in the genetic factors (i.e., the *genotype*) affecting the development of the characteristic?

The genetic factors are completely laid down when the parental sperm and ovum unite. Thus the individual's genotype, by definition, is "fixed" at the moment of conception. Of course, different potentials of the genotype may be expressed at different times in the course of the individual's development. But beyond conception, whatever we observe or measure of the organism is a phenotype, and this, by definition, is *not* "fixed." The phenotype is a result of the organism's internal genetic mechanisms established at conception and all the physical and social influences that impinge on the organism throughout the course of its development. Intelligence is a phenotype, not a genotype, so the argument about whether or not intelligence is "fixed" is seen to be spurious.

The really interesting and important question, which can be empirically answered by the methods of quantitative genetics, is: what is the correlation between genotypes and phenotypes at any given point in development? For continuous or metrical characteristics such as height and intelligence, the correlation, of course, can assume any value between 0 and 1. The square of the correlation between genotype and phenotype is technically known as the *heritability* of the characteristic, a concept which is discussed more fully in a later section.

The Stability of Intelligence Measures. The second aspect of the issue of "fixed intelligence" concerns the stability of intelligence measurements throughout the course of the individual's development. Since intelligence test scores are not points on an absolute scale of measurement like height and weight, but only indicate the individual's relative standing with reference to a normative population, the question we must ask is: To what extent do individuals maintain their standing relative to one another in measured intelligence over the course of time? The answer is to be found in the correlation between intelligence test scores on a group of persons at two points in time. Bloom (1964) has reviewed the major studies of this question and the evidence shows considerable consistency.

In surveying all the correlations reported in the literature between intelligence measured on the same individuals at two points in time, I have worked out a simple formula that gives a "best fit" to all these data. The formula has the virtue of a simple mnemonic, being much easier to remember than all the tables of correlations reported in the literature and yet being capable of reproducing the correlations with a fair degree of accuracy.

$$\hat{r}_{12} = r_{tt} \sqrt{\frac{CA_1}{CA_2}} \tag{1}$$

where $\hat{r}_{12}$ = the estimated correlation between tests given at times 1 and 2.

r_{tt} = the equivalent-forms or immediate test-retest reliability of the test.

CA_1 = the subject's chronological age at the time of the first test.

CA_2 = the subject's chronological age at the time of the second test.

Limitation: The formula holds only up to the point where CA_2 is age 10, at which time the empirical value of r_{12} approaches an asymptote, showing no appreciable increase thereafter. Beyond age 10, regardless of the interval between tests, the obtained test-retest correlations fall in the range between the test's reliability and the square of the reliability (i.e., $r_{tt} > r_{12} > r^2_{tt}$). These simple generalizations are intended simply as a means of summarizing the mass of empirical findings. They accord with Bloom's conclusion, based on his thorough survey of the published evidence, that beyond age 8, correlations between repeated tests of general intelligence, corrected for unreliability of measurement, are between $+$ .90 and unity (Bloom, 1964, p. 61) .

What these findings mean is that the IQ is not constant, but, like all other developmental characteristics, is quite variable early in life and becomes increasingly stable throughout childhood. By age 4 or 5, the IQ correlates about .70 with IQ at age 17, which means that approximately half (i.e., the square of the correlation) of the variance in adult intelligence can be predicted as early as age 4 or 5. This fact that half the variance in adult intelligence can be accounted for by age 4 has led to the amazing and widespread, but unwarranted and fallacious, conclusion that persons develop 50 percent of their mature intelligence by age 4! This conclusion, of course, does not at all logically follow from just knowing the magnitude of the correlation. The correlation between *height* at age 4 and at age 17 is also about .70, but who would claim that the square of the correlation indicated the proportion of adult height attained by age 4? The absurdity of this non sequitur is displayed in the prediction it yields: the average 4 year old boy should grow up to be 6 ft. 7 in. tall by age 17!

Intelligence has about the same degree of stability as other developmental characteristics. For example, up to age 5 or 6, height is somewhat more stable than intelligence, and thereafter the developmental rates of height and intelligence are about equally stable, except for a period of 3 or 4 years immediately after the onset of puberty, during which height is markedly less stable than intelligence. Intelligence is somewhat more stable than total body weight over the age range from 2 to 18 years. Intelligence has a considerably more stable growth rate than measures of physical strength (Bloom, 1964, pp. 46-47). Thus, although the IQ is certainly not "constant," it seems safe to say that under normal environmental conditions it is at least as stable as developmental characteristics of a strictly physical nature.

Intelligence as a Component of Mental Ability

The term "intelligence" should be reserved for the rather specific meaning I have assigned to it, namely, the general factor common to standard tests of intelligence. Any one verbal definition of this factor is really inadequate, but, if we must define it in so many words, it is probably best thought of as a capacity for abstract reasoning and problem solving.

What I want to emphasize most, however, is that *intelligence* should not be regarded as completely synonymous with what I shall call *mental ability*, a term which refers to the totality of a person's mental capabilities. Psychologists know full well that what they mean by intelligence in the technical sense is only a part of the whole spectrum of human abilities. The notion that a person's intelligence, or some test measurement thereof, reflects the totality of all that he can possibly do with his "brains" has long caused much misunderstanding and needless dispute. As I have already indicated, the particular constellation of abilities we now call "intelligence," and which we can measure by means of "intelligence" tests, has been singled out from the total galaxy of mental abilities as being especially important in our society mainly because of the nature of our traditional system of formal education and the occupational structure with which it is coordinated. Thus, the predominant importance of intelligence is derived, not from any absolute criteria or God-given desiderata, but from societal demands. But neither does this mean, as some persons would like to believe, that intelligence exists only "by definition" or is merely an insubstantial figment of psychological theory and test construction. Intelligence fully meets the usual scientific criteria for being regarded as an aspect of objective reality, just as much as do atoms, genes, and electromagnetic fields. Intelligence has indeed been singled out as especially important by the educational and occupational demands prevailing in all industrial societies, but it is nevertheless a biological reality and not just a figment of social convention. Where educators and society in general are most apt to go wrong is in failing fully to recognize and fully to utilize a broader spectrum of abilities than just that portion which psychologists have technically designated as "intelligence." But keep in mind that it is this technical meaning of "intelligence" to which the term specifically refers throughout the present article.

The Distribution of Intelligence

Intelligence tests yield numerical scores or IQs (intelligence quotients) which are assumed to be, and in fact nearly are, "normally" distributed in the population. That is, the distribution of IQs conforms to the normal or so-called Gaussian distribution, the familiar "bell-shaped curve." The IQ, which is now the most universal "unit" in the measurement of intelligence, was originally defined as the ratio of the individual's mental age (MA) to his chronological age (CA): IQ = (MA/CA) × 100. (Beyond about 16 years of age, the formula ceases to make sense.) Mental age was simply defined as the typical or average score obtained on a test by children of a given age, and thus the average child by definition has an IQ of 100. Because of certain difficulties with the mental age concept, which we need not go into here, modern test constructors no longer attempt to measure mental age but instead convert raw scores (i.e., the number of test items gotten "right") directly into IQs for each chronological age group. The average IQ at each age is arbitrarily set at 100, and the IQ is defined as a normally distributed variable with a mean of 100 and a standard deviation of 15 points. (The standard deviation is an index of the amount of dispersion of scores; in the normal distribution 99.7 percent of the scores fall within ± 3 standard deviations [i.e., ± 45 IQ points] of the mean.)

There is really nothing mysterious about the fact that IQs are "normally" distributed, but it is not quite sufficient, either, to say that the normality of the distribution is just an artifact of test construction. There is a bit more to it than that.

Toss a hundred or so pennies into the air and record the number of heads that come "up" when they fall. Do this several thousand times and plot a frequency distribution of the number of heads that come up on each of the thousands of throws. You will have a distribution that very closely approximates the normal curve, and the more times you toss the hundred pennies the closer you will approximate the normal distribution.

Now, a psychological test made up of 100 or so items would behave in the same manner as the pennies, and produce a perfectly normal distribution of scores, if (*a*) the items have an average difficulty level of 1/2 [i.e., exactly half of the number of persons taking the test would get the item "right"], and (*b*) the items are independent, that is, all the interitem correlations are zero. Needless to say, no psychological test that has ever been constructed meets these "ideal" criteria, and this is just as well, for if we succeeded in devising such a test it would "measure" absolutely nothing but chance variation. If the test is intended to measure some trait, such as general intelligence, it will be impossible for all the test items to be completely uncorrelated. They will necessarily have some degree of positive correlation among them. Then, if the items are correlated, and if we still want the test to spread people out over a considerable range of scores, we can achieve this only if the items vary in level of difficulty; they cannot all have a difficulty level of 1/2. (Imagine the extreme case in which all item intercorrelations were perfect and the difficulty level of all items was 1/2. Then the "distribution" of scores would have only two points: half the testees would obtain a score of zero and half would obtain a perfect score.) So we need to have test items which have an *average* difficulty level of 1/2 in the test overall, but which cover a considerable range of difficulty levels, say, from .1 to .9. Thus, test constructors make up their tests of items which have rather low average intercorrelations (usually between .1 and .2) and a considerable range of difficulty levels. These two sets of conditions working together, then, yield a distribution of test scores in the population which is very close to "normal." So far it appears as though we have simply made our tests in such a way as to *force* the scores to assume a normal distribution. And that is exactly true.

But the important question still remains to be answered: is intelligence itself—not just our measurements of it—really normally distributed? In this form the question is operationally meaningless, since, in order to find the form of the distribution of intelligence, we first have to measure it, and we have constructed our measuring instruments in such a way as to yield a normal distribution. The argument about the distribution of intelligence thus appears to be circular. Is there any way out? The only way I know of is to look for evidence that our intelligence scales or IQs behave like an "interval scale." On an interval scale, the interval between any two points is equal to the interval between any other two points the same numerical distance apart. Thus, intervals on the scale are equal and additive. If we *assume* that intelligence is "really" normally distributed in the population, and then measure it in such a way that we obtain a normal distribution of scores, our measurements (IQs) can be regarded as constituting an interval scale. If, then, the scale in fact behaves like an interval scale, there is some justification for saying that intelligence itself (not just IQ) is normally distributed. What evidence is there of the IQ's behaving like an interval scale? The most compelling evidence, I believe, comes from studies of the inheritance of intelligence, in which we examine the pattern of intercorrelations among relatives of varying degrees of kinship.

But, first, to understand what is meant by "behaving" like an interval scale, let us look at two well-known interval scales, the Fahrenheit and centigrade thermometers. We can prove that these are true interval scales by showing that they "behave" like interval scales in the following manner: Mix a pint of ice water at $0°$ C with a pint of boiling water at $100°$ C. The resultant temperature of the mixture will be $50°$ C. Mix 3 pints of ice water with 1 pint of boiling water and the temperature of the mix will be $25°$ C. And we can continue in this way, mixing various proportions of water at different temperatures and predicting the resultant temperatures on the assumption of an interval scale. To the extent that the thermometer readings fit the predictions, they can be considered an interval scale.

Physical stature (height) is measured on an interval scale (more than that, it is also a ratio scale) in units which are independent of height, so the normal distribution of height in the population is clearly a fact of nature and not an artifact of the scale of measurement. A rather simple genetic model "explains" the distribution of height by hypothesizing that individual variations in height are the result of a large number of independent factors each having a small effect in determining stature. (Recall the penny-tossing analogy.) This model predicts quite precisely the amount of "regression to the population mean" of the children's average height from the parent's average height, a phenomenon first noted by Sir Francis Galton in 1885. The amount of "regression to the mean" from grandparent to grandchild is exactly double that from parent to child. These regression lines for various degrees of kinship are perfectly rectilinear throughout the entire range, except at the very lower end of the scale of height, where one finds midgets and dwarfs. The slope of the regression line changes in discrete jumps according to the remoteness of kinship of the groups being compared. All this could happen only if height were measured on an interval scale. The regression lines would not be rectilinear if the trait (height) were not measured in equal intervals.

Now, it is interesting that intelligence measurements show about the same degree of "filial regression," as Galton called it, that we find for height. The simple polygenic model for the inheritance of height fits the kinship correlations obtained for intelligence almost as precisely as it does for height. And the kinship regression lines are as rectilinear for intelligence as for height, throughout the IQ scale, except at the very lower end, where we find pathological types of mental deficiency analogous to midgets and dwarfs on the scale of physical stature. In brief, IQs behave just about as much like an interval scale as do measurements of height, which we know for sure is an interval scale. Therefore, it is not unreasonable to treat the IQ as an interval scale.

Although standardized tests such as the Stanford-Binet and the Wechsler Scales were each constructed by somewhat different approaches to achieving interval scales, they both agree in revealing certain systematic discrepancies from a perfectly normal distribution of IQs when the tests are administered to a very large and truly random sample of the population. These slight deviations of the distribution of IQs from perfect normality have shown up in many studies using a variety of tests. The most thorough studies and sophisticated discussions of their significance can be found in articles by Sir Cyril Burt (1957, 1963). The evidence, in short, indicates that intelligence is *not* distributed quite normally in the population. The distribution of IQs approximates normality quite closely in the IQ range from about 70 to 130. But outside this range there are slight, although very significant, departures from normality. From a scientific standpoint, these discrepancies are of considerable interest as genuine phenomena needing explanation.

Figure I shows an idealized distribution of IQs if they were distributed perfectly normally. Between IQ 70 and IQ 130, the percentage of cases falling between different IQ intervals, as indicated in Figure 1, are very close to the actual percentages estimated from large samples of the population and the departures are hardly enough to matter from any practical standpoint.

Examination of this normal curve can be instructive if one notes the consequences of shifting the total distribution curve up or down the IQ scale. The consequences of a given shift become more extreme out toward the "tails" of the distribution. For example, shifting the mean of the distribution from 100 down to 90 would put 50 percent instead of only 25 percent of the population below IQ 90; and it would put 9 percent instead of 2 percent below IQ 70. And in the upper tail of the distribution, of course, the consequences would be the reverse; instead of 25 percent above IQ 110, there would be only 9 percent, and so on. The point is that relatively small shifts in the mean of the IQ distribution can result in very large differences in the proportions of the population that fall into

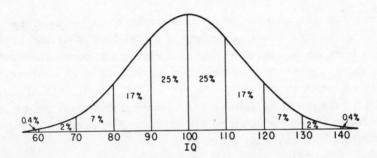

FIGURE 1.

The theoretical normal or Gaussian distribution of IQs, showing the expected percentages of the population in each IQ range. Except at the extremes (below 70 and above 130) these percentages are very close to actual population values. (The percentage figures total slightly more than 100% because of rounding.)

the very low or the very high ranges of intelligence. A 10 point downward shift in the mean, for example, would more than triple the percentage of mentally retarded (IQs below 70) in the population and would reduce the percentage of intellectually "gifted" (IQs above 130) to less than one-sixth of their present number. It is in these tails of the normal distribution that differences become most conspicuous between various groups in the population that show mean IQ differences, for whatever reason, of only a few IQ points. From a knowledge of relatively slight mean differences between various social class and ethnic groups, for example, one can estimate quite closely the relatively large differences in their proportions in special classes for the educationally retarded and for the "gifted" and in the percentages of different groups receiving scholastic honors at graduation. It is simply a property of the normal distribution that the effects of group differences in the mean are greatly magnified in the different proportions of each group that we find as we move further out toward the upper or lower extremes of the distribution.

I indicated previously that the distribution of intelligence is really not quite "normal," but shows certain systematic departures from "normality." These departures from the normal distribution are shown in Figure 2 in a slightly exaggerated form to make them clear. The shaded area is the normal distribution; the heavy line indicates the actual distribution of IQs in the population. We note that there are more very low IQs than would be expected in a truly normal distribution,

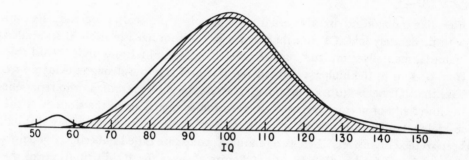

FIGURE 2.

Theoretical "normal" distribution of IQs (shaded curve) and the actual distribution in the population (heavy line), with the lower hump exaggerated for explanatory purposes. See text for explanation.

and also there is an excess of IQs at the upper end of the scale. Note, too, the slight excess in the IQ range between about 70 and 90.

The very lowest IQs, below 55 or 60, we now know, really represent a different distribution from that of the rest of the intelligence distribution (Roberts, 1952; Zigler, 1967). Whatever factors are responsible for individual differences in the IQ range above 60 are not sufficient to account for IQs below this level, and especially below IQ 50. Practically all IQs below this level represent severe mental deficiency due to pathological conditions, massive brain damage, or rare genetic and chromosomal abnormalities. Only about 1/2 to 3/4 of 1 percent of the total population falls into the IQ range below 50; this is fewer than 1/3 of all individuals classed as mentally retarded (IQs below 70). These severe grades of mental defect are not just the lower extreme of normal variation. Often they are due to a single recessive or mutant gene whose effects completely override all the other genetic factors involved in intelligence; thus they have been called "major gene" defects. In this respect, the distribution of intelligence is directly analogous to the distribution of stature. Short persons are no more abnormal than are average or tall persons; all are instances of normal variation. But extremely short persons at the very lower end of the distribution are really part of another, abnormal, distribution, generally consisting of midgets and dwarfs. They are clearly not a part of normal variation. One of the commonest types of dwarfism, for example, is known to be caused by a single recessive gene.

Persons with low IQs caused by major gene defects or chromosomal abnormalities, like mongolism, are also usually abnormal in physical appearance. Persons with moderately low IQs that represent a part of normal variation, the so-called "familial mentally retarded," on the other hand, are physically indistinguishable from persons in the higher ranges of IQ. But probably the strongest evidence we have that IQs below 50 are a group apart from the mildly retarded, who represent the lower end of normal variation, comes from comparisons of the siblings of the severely retarded with siblings of the mildly retarded. In England, where this has been studied intensively, these two retardate groups are called imbecile (IQs below 50) and feebleminded (IQs 50 to 75). Figure 3 shows the IQ distributions of the *siblings* of imbecile and feebleminded children (Roberts, 1952). Note that the siblings of imbeciles have a much higher average level of intelligence than the siblings of the feebleminded. The latter group, furthermore, shows a distribution of IQs that would be predicted from a genetic model intended to account for the normal variation of IQ in the population. This model does not at all predict the IQ distribution for the imbecile sibships. To explain the results shown in Figure 3 one must postulate some additional factors (gene or chromosome defects, pathological conditions, etc.) that cause imbecile and idiot grades of mental deficiency.

Another interesting point of contrast between severe mental deficiency and mild retardation is the fact noted by Kushlick (1966, p. 130), in surveying numerous studies, that "The parents of severely subnormal children are evenly distributed among all the social strata of industrial society, while those of mildly subnormal subjects come predominantly from the lower social classes. There is

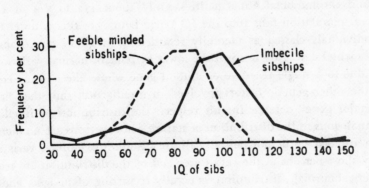

FIGURE 3.

Frequency distributions of the IQs of sibs of feebleminded and imbeciles of the IQ range 30-68. (Roberts, 1952.)

now evidence which suggests that mild subnormality in the absence of abnormal neurological signs (epilepsy, electroencephalographic abnormalities, biochemical abnormalities, chromosomal abnormalities or sensory defects) is virtually confined to the lower social classes. Indeed, there is evidence that almost no children of higher social class parents have IQ scores of less than 80, unless they have one of the pathological processes mentioned above."

In the remainder of this article we shall not be further concerned with these exceptionally low IQs below 50 or 60, which largely constitute a distribution of abnormal conditions superimposed on the factors that make for normal variation in intelligence. We shall be mainly concerned with the factors involved in the normal distribution.

Returning to Figure 2, the best explanation we have for the "bulge" between 70 and 90 is the combined effects of severe environmental disadvantages and of emotional disturbances that depress test scores. Burt (1963) has found that when, independent of the subjects' test performance there is evidence for the existence of factors that depress performance, and these exceptional subjects' scores are removed from the distribution, this "bulge" in the 70-90 range is diminished or erased. Also, on retest under more favorable conditions, the IQs of many of these exceptional subjects are redistributed at various higher points on the scale, thereby making the IQ distribution more normal.

The "excess" of IQs at the high end of the scale is certainly a substantial phenomenon, but it has not yet been adequately accounted for. In his multifactorial theory of the inheritance of intelligence, Burt (1958) has postulated major gene effects that make for exceptional intellectual abilities represented at the upper end of the scale, just as other major gene effects make for the subnormality found at the extreme lower end of the scale. One might also hypothesize that superior genotypes for intellectual development are pushed to still greater superiority in their phenotypic expression through interaction with the environment. Early recognition of superiority leads to its greater cultivation and encouragement by the individual's social environment. This influence is keenly evident in the developmental histories of persons who have achieved exceptional eminence (Goertzel & Goertzel, 1962). Still another possible explanation of the upper-end "excess" lies in the effects of assortative mating in the population, meaning the tendency for "like to marry like." If the degree of resemblance in intelligence between parents in the upper half of the IQ distribution were significantly greater than the degree of resemblance of parents in the below-average range, genetic theory would predict the relative elongation of the upper tail of the distribution.

This explanation, however, must remain speculative until we have more definite evidence of whether there is differential assortative mating in different regions of the IQ distribution.

The Concept of Variance. Before going on to discuss the factors that account for normal variation in intelligence among individuals in the population, a word of explanation is in order concerning the quantification of variation. The amount of dispersion of scores depicted by the distributions in Figures 1 and 2 is technically expressed as the *variance,* which is the square of the standard deviation of the scores in the distribution. (Since the standard deviation of IQs in the population is 15, the total variance is 225.) *Variance* is a basic concept in all discussions of individual differences and population genetics. If you take the difference between every score and the mean of the total distribution, square each of these differences, sum them up, and divide the sum by the total number of scores, you have a quantity called the *variance*. It is an index of the total amount of variation among scores. Since variance represents variation on an additive scale, the total variance of a distribution of scores can be partitioned into a number of components, each one due to some factor which contributes a certain specifiable proportion of the variance, and all these variance components add up to the total variance. The mathematical technique for doing this, called "the analysis of variance," was invented by Sir Ronald Fisher, the British geneticist and statistician. It is one of the great achievements in the development of statistical methodology.

The Inheritance of Intelligence

"In the actual race of life, which is not to get ahead, but to get ahead of somebody, the chief determining factor is heredity." So said Edward L. Thorndike in 1905. Since then, the preponderance of evidence has proved him right, certainly as concerns those aspects of life in which intelligence plays an important part.

But one would get a quite different impression from reading most of the recent popular textbooks of psychology and education. Genetic factors in individual differences have usually been belittled, obscured, or denigrated, probably for reasons of interest mainly on historical, political, and ideological grounds which we need not go into here. Some of the following quotations, each from different widely used texts in our field, give some indication of the basis for my complaint. "We can attribute no particular portion of intelligence to heredity and no particular portion to the environment." "The relative influence of heredity and environment upon intelligence has been the topic of considerable investigations over the last half century. Actually the problem is incapable of solution since studies do not touch upon the problem of heredity and environment but simply upon the susceptibility of the content of a particular test to environmental influences." "Among people considered normal, the range of genetic variations is not very great." "Although at the present time practically all responsible workers in the field recognize that conclusive proof of the heritability of mental ability (where no organic or metabolic pathology is involved) is still lacking, the assumption that subnormality has a genetic basis continues to crop up in scientific studies." "There is no evidence that nature is more important than nurture. These two forces always operate together to determine the course of intellectual development." The import of such statements apparently filters up to high levels of policy-making, for we find a Commissioner of the U.S. Office of Education stating in a published speech that children ". . . all have similar potential at birth. The differences occur shortly thereafter." These quotations typify much of the current attitude toward heredity and environment that has prevailed in education in recent years. The belief in the almost infinite plasticity of intellect, the ostrich-like denial of biological factors in individual differences, and the slighting of the role of genetics in the study of intelligence can only hinder investigation and understanding of the conditions, processes, and limits through which the social environment influences human behavior.

But fortunately we are beginning to see some definite signs that this mistreatment of the genetic basis of intelligence by social scientists may be on the wane, and that a biosocial view of intellectual development more in accord with the evidence is gaining greater recognition. As Yale psychologist Edward Zigler (1968) has so well stated:

Not only do I insist that we take the biological integrity of the organism seriously, but it is also my considered opinion that our nation has more to fear from unbridled environmentalists than they do from those who point to such integrity as one factor in the determination of development. It is the environmentalists who have been writing review after review in which genetics are ignored and the concept of capacity is treated as a dirty word. It is the environmentalists who have placed on the defensive any thinker who, perhaps impressed by the revolution in biological thought stemming from discoveries involving RNA-DNA phenomena, has had the temerity to suggest that certain behaviors may be in part the product of read-out mechanisms residing within the programmed organism. It is the unbridled environmentalist who emphasizes the plasticity of the intellect, that tells us one can change both the general rate of development and the configuration of intellectual processes which can be referred to as the intellect, if we could only subject human beings to the proper technologies. In the educational realm, this has spelled itself out in the use of panaceas, gadgets, and gimmicks of the most questionable sort. It is the environmentalist who suggests to parents how easy it is to raise the child's IQ and who has prematurely led many to believe that the retarded could be made normal, and the normal made geniuses. It is the environmentalist who has argued for pressure-cooker schools, at what psychological cost, we do not yet know.

Most geneticists and students of human evolution have fully recognized the role of culture in shaping "human nature," but also they do not minimize the biological basis of diversity in human behavioral characteristics. Geneticist Theodosius Dobzhansky (1968, p. 554) has expressed this viewpoint in the broadest terms: "The trend of cultural evolution has been not toward making everybody have identical occupations but toward a more and more differentiated occupational structure. What would be the most adaptive response to this trend? Certainly nothing that would encourage genetic uniformity.... To argue that only environmental circumstances and training determine a person's behavior makes a travesty of democratic notions of individual choice, responsibility, and freedom."

Evidence from Studies of Selective Breeding

The many studies of selective breeding in various species of mammals provide conclusive evidence that many behavioral characteristics, just as most physical characteristics, can be manipulated by genetic selection (see Fuller & Thompson, 1962; Scott and Fuller, 1965). Rats, for example, have been bred for maze learning ability in many different laboratories. It makes little difference whether one refers to this ability as rat "intelligence," "learning ability" or some other term— we know that it is possible to breed selectively for whatever the factors are that make for speed of maze learning. To be sure, individual variation in this complex ability may be due to any combination of a number of characteristics involving sensory acuity, drive level, emotional stability, strength of innate turning preferences, brain chemistry, brain size, structure of neural connections, speed of synaptic transmission, or whatever. The point is that the molar behavior of learning to get through a maze efficiently without making errors (i.e., going up blind alleys) can be markedly influenced in later generations by selective breeding of the parent generations of rats who are either fast or slow ("maze bright" or "maze dull," to use the prevailing terminology in this research) in learning to get through the maze. Figure 4 shows the results of one such genetic selection experiment. They are quite typical; within only six generations

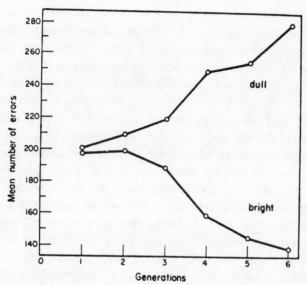

FIGURE 4.

The mean error scores in maze learning for successive generations of selectively bred "bright" and "dull" strains of McGill rats. (After Thompson, 1954.)

of selection the offspring of the "dull" strain make 100 percent more errors in learning the maze than do the offspring of the "bright" strain (Thompson, 1954). In most experiments of this type, of course, the behaviors that respond so dramatically to selection are relatively simple as compared with human intelligence, and the experimental selection pressure is severe, so the implications of such findings for the study of human variation should not be overdrawn. Yet geneticists seem to express little doubt that many behavioral traits in humans would respond similarly to genetic selection. Three eminent geneticists (James F. Crow, James V. Neel, and Curt Stern) of the National Academy of Sciences recently prepared a "position statement," which was generally hedged by extreme caution and understatement, that asserted: "Animal experiments have shown that almost any trait can be changed by selection. ... A selection program to increase human intelligence (or whatever is measured by various kinds of 'intelligence' tests) would almost certainly be successful in some measure. The same is probably true for other behavioral traits. The *rate* of increase would be somewhat unpredictable, but there is little doubt that there would be progress" (National Academy of Sciences, 1967, p. 893).

Direct Evidence of Genetic Influences on Human Abilities

One of the most striking pieces of evidence for the genetic control of mental abilities is a chromosomal anomaly called Turner's syndrome. Normal persons have 46 chromosomes. Persons with Turner's syndrome have only 45. When their chromosomes are stained and viewed under the microscope, it is seen that the sex-chromatin is missing from one of the two chromosomes that determine the individual's sex. In normal persons this pair of chromosomes is conventionally designated XY for males and XX for females. The anomaly of Turner's syndrome is characterized as XO. These persons always have the morphologic appearance of females but are always sterile, and they show certain physical characteristics such as diminutive stature, averaging about five feet tall as adults. The interesting point about Turner's cases from our standpoint is that although their IQs on most verbal tests of intelligence show a perfectly normal distribution, their performance on tests involving spatial ability or perceptual organization is abnormally low (Money, 1964). Their peculiar deficiency in spatial-perceptual ability is sometimes so severe as to be popularly characterized as "space-form blindness." It is also interesting that Turner's cases seem to be more or less uniformly low on spatial ability regardless of their level of performance on other tests of mental ability. These rare persons also report unusual difficulty with arithmetic and mathematics in school despite otherwise normal or superior intelligence. So here is a genetic aberration, clearly identifiable under the microscope, which has quite specific consequences on cognitive processes. Such specific intellectual deficiencies are thus entirely possible without there being any specific environmental deprivations needed to account for them.

There are probably other more subtle cognitive effects associated with the sex chromosomes in normal persons. It has long been suspected that males have greater environmental vulnerability than females, and Nancy Bayley's important longitudinal research on children's mental development clearly shows both a higher degree and a greater variety of environmental and personality correlates of mental abilities in boys than in girls (Bayley, 1965b, 1966, 1968).

Polygenic Inheritance

Since intelligence is basically dependent on the structural and biochemical properties of the brain, it should not be surprising that differences in intellectual capacity are partly the result of genetic factors which conform to the same principles involved in the inheritance of physical characteristics. The general model that geneticists have devised to account for the facts of inheritance of continuous or metrical physical traits, such as stature, cephalic index, and finger-print ridges, also applies to intelligence. *The mechanism of inheritance for such traits is called polygenic, since normal variation in the characteristic is the result of multiple genes whose effects are small, similar, and cumulative.* The genes can be thought of as the pennies in the coin-tossing analogy described pre-viously. Some genes add a positive increment to the metric value of the charac-teristic ("heads") and some genes add nothing ("tails"). The random segrega-tion of the parental genes in the process of gametogenesis (formation of the sex cells) and their chance combination in the zygote (fertilized egg) may be likened to the tossing of a large number of pennies, with each "head" adding a positive increment to the trait, thereby producing the normal bell-shaped distribution of trait values in a large number of tosses. The actual number of genes involved in intelligence is not known. In fact, the total number of genes in the human chromosomes is unknown. The simplest possible model would require between ten and twenty gene pairs (alleles) to account for the normal distribution of intelligence, but many more genes than this are most likely involved (Gottesman, 1963, pp. 290-291).

The Concept of Heritability

The study of the genetic basis of individual differences in intelligence in humans has evolved in the traditions and methods of that branch of genetics called quantitative genetics or population genetics, the foundations of which were laid down by British geneticists and statisticians such as Galton, Pearson, Fisher, Haldane, and Mather, and, in the United States, by J. L. Lush and Sewall Wright. Probably the most distinguished exponent of the application of these methods to the study of intelligence is Sir Cyril Burt, whose major writings on this subject are a "must" for students of individual differences (Burt, 1955, 1958, 1959, 1961, 1966; Burt & Howard, 1956, 1957).

One aim of this approach to the study of individual differences in intelligence is to account for the total variance in the population (excluding pathological cases at the bottom of the distribution) in terms of the proportions of the variance attributable to various genetic and environmental components. It will pay to be quite explicit about just what this actually means.

Individual differences in such measurements of intelligence as the IQ are represented as population variance in a phenotype V_P, and are distributed approximately as shown in Figure 1. Conceptually, this total variance of the phenotypes can be partitioned into a number of variance components, each of which represents a source of variance. The components, of course, all add up to the total variance. Thus,

$$V_P = \frac{\overbrace{(V_G + V_{AM}) + V_D + V_i}^{V_H}}{} + \frac{\overbrace{V_E + 2\,Cov_{HE} + V_I}^{V_E}}{} + \frac{V_e}{} \qquad (2)$$

$$\underbrace{\phantom{\frac{(V_G + V_{AM}) + V_D + V_i}{V_H}}}_{\text{Heredity}} \quad \underbrace{\phantom{\frac{V_E + 2\,Cov_{HE} + V_I}{V_E}}}_{\text{Environment}} \quad \underbrace{\phantom{\frac{V_e}{}}}_{\text{Error}}$$

where: V_P = phenotypic variance in the population

V_G = genic (or additive) variance

V_{AM} = variance due to assortative mating. $V_{AM} = O$ under random mating (panmixia).

V_D = dominance deviation variance

V_i = epistatis (interaction among genes at 2 or more loci)

V_E = environmental variance

Cov_{HE} = covariance of heredity and environment

V_I = true statistical interaction of genetic and environmental factors

V_e = error of measurement (unreliability).

Here are a few words of explanation about each of these variance components.

Phenotypic Variance. V_P is already clear; it is the total variance of the trait measurements in the population.

Genic Variance. V_G, the genic (or additive) variance, is attributable to gene effects which are additive; that is, each gene adds an equal increment to the metric value of the trait. Sir Ronald Fisher referred to this component as "the essential genotypes," since it is the part of the genetic inheritance which "breeds true"—it accounts for the resemblance between parents and offspring. If trait variance involved nothing but additive genic effects, the average value of all the offspring that could theoretically be born to a pair of parents would be exactly equal to the average value of the parents (called the midparent value). It is thus the genic aspect which is most important to agriculturalists and breeders of livestock, since it is the genic component of the phenotypic variance that responds to selection according to the simple rule of "like begets like." The larger the proportion of genic variance involved in a given characteristic, the fewer is the number of generations of selective breeding required to effect a change of some specified magnitude in the characteristic.

Assortative Mating. V_{AM}, the variance due to assortative mating, is conventionally not separated from V_G, since assortative mating actually affects the proportion of V_G directly. I have separated these components here for explanatory reasons, and it is, in fact, possible to obtain independent estimates of the two components. If mating were completely random in the population with respect to a given characteristic—that is, if the correlation between parents were zero (a state of affairs known as *panmixia*)—the V_{AM} component would also be equal to zero and the population variance on the trait in question would therefore be reduced.

Assortative mating has the effect of increasing the resemblance among siblings and also of increasing the differences between families in the population. (In the terminology of analysis of variance, assortative mating decreases the *Within* Families variance and increases the *Between* Families variance.)

For some human characteristics the degree of assortative mating is effectively zero. This is true of fingerprint ridges, for example. Men and women are obviously not attracted to one another on the basis of their fingerprints. Height, however, has an assortative mating coefficient (i.e., the correlation between mates) of about .30. The IQ, interestingly enough, shows a higher degree of assortative mating in our society than any other measurable human characteristic. I have surveyed the literature on this point, based on studies in Europe and North America, and find that the correlation between spouses' intelligence test scores averages close to $+.60$. Thus, spouses are more alike in intelligence than brothers and sisters, who are correlated about .50.

As Eckland (1967) has pointed out, this high correlation between marriage partners does not come about solely because men and women are such excellent judges of one another's intelligence, but because mate selection is greatly aided by the highly visible selective processes of the educational system and the occupational hierarchy. Here is a striking instance of how educational and social factors can have far-reaching genetic consequences in the population. One would predict, for example, that in preliterate or preindustrial societies assortative mating with respect to intelligence would be markedly less than it is in modern industrial societies. The educational screening mechanisms and socioeconomic stratification by which intelligence becomes more readily visible would not exist, and other traits of more visible importance to the society would take precedence over intelligence as a basis for assortative mating. Even in our own society, there may well be differential degrees of assortative mating in different segments of the population, probably related to their opportunities for educational and occupational selection. When any large and socially insulated group is not subject to the social and educational circumstances that lead to a high degree of assortative mating for intelligence, there should be important genetic consequences. One possible consequence is some reduction of the group's ability, not as individuals but as a group, to compete intellectually. Thus probably one of the most cogent arguments for society's promoting full equality of educational, occupational, and economic opportunity lies in the possible genetic consequences of these social institutions.

The reason is simply that assortative mating increases the genetic variance in the population. By itself this will not affect the mean of the trait in the population, but it will have a great effect on the proportion of the population falling in the upper and lower tails of the distribution. Under present conditions, with an assortative mating coefficient of about .60, the standard deviation of IQs is 15 points. If assortative mating for intelligence were reduced to zero, the standard deviation of IQs would fall to 12.9. The consequences of this reduction in the standard deviation would be most evident at the extremes of the intelligence distribution. For example, assuming a normal distribution of IQs and the present standard deviation of 15, the frequency (per million) of persons above IQ 130 is 22,750. Without assortative mating the frequency of IQs over 130 would fall to 9,900, or only 43.5 percent of the present frequency. For IQs above 145, the frequency (per million) is 1,350 and with no assortative mating would fall to 241, or 17.9 percent of the present frequency. And there are now approximately 20 times as many persons above an IQ of 160 as we would find if there were no assortative mating for intelligence.[3] Thus differences in assortative mating can have a profound effect on a people's intellectual resources, especially at the levels of intelligence required for complex problem solving, invention, and scientific and technological innovation.

But what is the effect of assortative mating on the lower tail of the distribution? On theoretical grounds we should also expect it to increase the proportion of low IQs in the population. It probably does this to some extent, but not as much as it increases the frequency of higher IQs, because there is a longer-term consequence of assortative mating which must also be considered. A number of studies have shown that in populations practicing a high degree of assortative mating, persons below IQ 75 are much less successful in finding marriage partners and, as a group, have relatively fewer offspring than do persons of higher intelligence (Bajema, 1963, 1966; Higgins, Reed, & Reed, 1962). Since assortative mating increases variance, it in effect pushes more people into the below IQ

---

[3] I am grateful to University of California geneticist Dr. Jack Lester King for making these calculations, which are based on the assumption that the heritability of IQ is .80, a value which is the average of all the major studies of the heritability of intelligence.

75 group, where they fail to reproduce, thereby resulting in a net selection for genes favoring high intelligence. Thus, in the long run, assortative mating may have a eugenic effect in improving the general level of intelligence in the population.

Dominance Deviation. V_D, the dominance deviation variance, is apparent when we observe a systematic discrepancy between the average value of the parents and the average value of their offspring on a given characteristic. Genes at some of the loci in the chromosome are recessive (r) and their effects are not manifested in the phenotype unless they are paired with another recessive at the same locus. If paired with a dominant gene (D), their effect is overridden or "dominated" by the dominant gene. Thus, in terms of increments which genes add to the metric value of the phenotype, if $r = 0$ and $D = 1$, then $r + r = 0$, and $D + D = 2$, but $D + r$ will equal 2, since D dominates r. Because of the presence of some proportion of recessive genes in the genotypes for a particular trait, not all of the parents' phenotypic characteristics will show up in their offspring, and, of course, vice versa: not all of the offspring's characteristics will be seen in the parents. This makes for a less than perfect correlation between midparent and midchild values on the trait in question. V_D, the dominance variance, represents the component of variance in the population which is due to this average discrepancy between parents and offspring. The magnitude of V_D depends upon the proportions of dominant and recessive genes constituting the genotypes for the characteristic in the population.

Epistasis. V_I is the variance component attributable to epistasis, which means the interaction of the effects among genes at two or more loci. When genes "interact," their effects are not strictly additive; that is to say, their combined effect may be more or less than the sum of their separate effects. Like dominance, epistasis also accounts for some of the lack of resemblance between parents and their offspring. And it increases the population variance by a component designated as V_I.

Environmental Variance. "Environmental" really means all sources of variance not attributable to genetic effects or errors of measurement (i.e., test unreliability). In discussions of intelligence, the environment is often thought of only in terms of the social and cultural influences on the individual. While these are important, they are not the whole of "environment," which includes other more strictly biological influences, such as the prenatal environment and nutritional factors early in life. In most studies of the heritability of intelligence "environment" refers to all variance that is not accounted for by genetic factors [$(V_G + V_{AM}) + V_D + V_1$] and measurement error (V_e).

Covariance of Heredity and Environment. This term can also be expressed as $2r_{HE} \sqrt{V_H \times V_E}$, where r_{HE} is the correlation between heredity and environment, V_H is the variance due to all genetic factors, and V_E is variance due to all environmental factors. In other words, if there is a positive correlation between genetic and environmental factors, the population variance is increased by a theoretically specifiable amount indicated by the covariance term in Equation 2.

Such covariance undoubtedly exists for intelligence in our society. Children with better than average genetic endowment for intelligence have a greater than chance likelihood of having parents of better than average intelligence who are capable of providing environmental advantages that foster intellectual development. Even among children within the same family, parents and teachers will often give special attention and opportunities to the child who displays exceptional abilities. A genotype for superior ability may cause the social environment to foster the ability, as when parents perceive unusual responsiveness to music in one of their children and therefore provide more opportunities for listening, music lessons, encouragement to practice, and so on. A bright child may also create a more intellectually stimulating environment for himself in terms of the kinds of activities that engage his interest and energy. And the social rewards that come to the individual who excels in some activity reinforce its further development. Thus the covariance term for any given trait will be affected to a significant degree by the kinds of behavioral propensities the culture rewards or punishes, encourages or discourages. For traits viewed as desirable in our culture, such as intelligence, hereditary and environmental factors will be positively correlated. But for some other traits which are generally viewed as socially undesirable, hereditary and environmental influences may be negatively correlated. This means that the social environment tends to discourage certain behavioral propensities when they are out of line with the values of the culture. Then, instead of heredity and environment acting in the same direction, they work in opposite directions, with a consequent reduction in the population variance in the trait. Overt aggressive tendencies may be a good example of behavior involving a negative correlation between genotypic propensities and environmental counter-pressures. An example of negative heredity-environment correlation in the scholastic realm would be found in the case where a child with a poor genetic endowment for learning some skill which is demanded by societal norms, such as being able to read, causes the child's parents to lavish special tutorial attention on their child in an effort to bring his performance up to par.

In making overall estimates of the proportions of variance attributable to hereditary and environmental factors, there is some question as to whether the covariance component should be included on the side of heredity or environment. But there can be no "correct" answer to this question. To the degree that the individual's genetic propensities cause him to fashion his own environment, given the opportunity, the covariance (or some part of it) can be justifiably regarded as part of the total heritability of the trait. But if one wishes to estimate what the heritability of the trait would be under artificial conditions in which there is absolutely no freedom for variation in individuals' utilization of their environment, then the covariance term should be included on the side of environment. Since most estimates of the heritability of intelligence are intended to reflect the existing state of affairs, they usually include the covariance in the proportion of variance due to heredity.

Interaction of Heredity and Environment. The *interaction* of genetic and environmental factors (V_I) must be clearly distinguished from the *covariance* of heredity and environment. There is considerable confusion concerning the meaning of interaction in much of the literature on heredity and intelligence. It is claimed, for example, that nothing can be said about the relative importance of heredity and environment because intelligence is the result of the "interaction" of these influences and therefore their independent effects cannot be estimated. This is simply false. The proportion of the population variance due to genetic $\times$ environment interaction is conceptually and empirically separable from other variance components, and its independent contribution to the total variance can be known. Those who call themselves "interactionists," with the conviction that they have thereby either solved or risen above the whole issue of the relative contributions of heredity and environment to individual differences in intelligence, are apparently unaware that the preponderance of evidence indicates that the interaction variance, V_I is the smallest component of the total phenotypic variance of intelligence.

What *interaction* really means is that different genotypes respond in different ways to the same environmental factors. For example, genetically different individuals having the same initial weight and the same activity level may gain

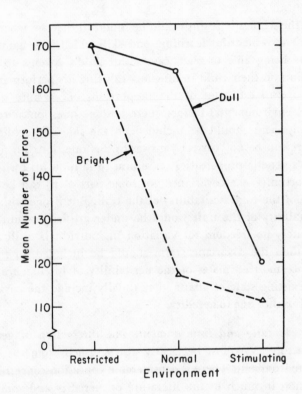

FIGURE 5.

Illustration of a true genotype × environment interaction for error scores in maze learning by "bright" and "dull" strains of rats raised in "restricted," "normal," and "stimulating" environments. (After Cooper & Zubek, 1958.)

weight at quite different rates all under exactly the same increase in caloric intake. Their genetically different constitutions cause them to metabolize exactly the same intake quite differently. An example of genotype × environmental interaction in the behavioral realm is illustrated in Figure 5. Strains of rats selectively bred for "brightness" or "dullness" in maze learning show marked differences in maze performance according to the degree of sensory stimulation in the conditions under which they are reared. For the "bright" strain, the difference between being reared in a "restricted" or in a "normal" environment makes a great difference in maze performance. But for the "dull" strain the big difference is between a "normal" and a "stimulating" environment. While the strains differ greatly when reared under "normal" conditions (presumably the conditions under which they were selectively bred for "dullness" and "brightness"), they do not differ in the least when reared in a "restricted" environment and only slightly in a "stimulating" environment. This is the meaning of the genetic $\times$ environment interaction. Criticisms of the analysis of variance model for the components of phenotypic variance (e.g., Equation 2), put forth first by Loevinger (1943) and then by Hunt (1961, p. 329), are based on the misconception that the model implies that all effects of heredity and environment are strictly additive and there is no "non-additive" or interaction term. The presence of V_I in Equation 2 explicitly shows that the heredity $\times$ environment interaction is included in the analysis of variance model, and the contribution of V_I to the total variance may be estimated independently of the purely additive effects of heredity and environment. The magnitude of V_I for any given characteristic in any specified population is a matter for empirical study, not philosophic debate. If V_I turns out to constitute a relatively small proportion of the total variance, as the evidence shows is the case for human intelligence, this is not a fault of the analysis of variance model. It is simply a fact. If the interaction variance actually exists in any significant amount, the model will reveal it.

Several studies, reviewed by Wiseman (1964, p. 55; 1966, p. 66), provide most of the information we have concerning what may be presumed to be an heredity $\times$ environment interaction with respect to human intelligence. The general finding is that children who are more than one standard deviation (SD) above the mean IQ show greater correlations with environmental factors than do children who are more than one SD below the mean. In other words, if the heritability of IQ were determined in these two groups separately, it would be higher in the low IQ groups. Also, when siblings within the same family are grouped into above and below IQ 100, the scholastic achievement of the above 100 group shows a markedly higher correlation with environmental factors than in the below 100 group. This indicates a true interaction between intelligence and environment in determining educational attainments.

Error Variance. The variance due to errors of measurement (V_e) is, of course, unwanted but unavoidable, since all measurements fall short of perfect reliability. The proportion of test score variance due to error is equal to $1 - r_{tt}$ (where r_{tt} is the reliability of the test, that is, its correlation with itself). For most intelligence tests, error accounts for between 5 and 10 percent of the variance.

Definition of Heritability

Heritability is a technical term in genetics meaning specifically the proportion of phenotypic variance due to variance in genotypes. When psychologists speak of heritability they almost invariably define it as:

$$H = \frac{(V_G + V_{AM}) + V_D + V_i}{V_P - V_e} \tag{3}$$

Although this formula is technically the definition of H, heritability estimates in psychological studies may also include the covariance term of Equation 2 in the numerator of Equation 3.

Common Misconceptions About Heritability

Certain misconceptions about heritability have become so widespread and strongly ingrained that it is always necessary to counteract them before presenting the empirical findings on the subject, lest these findings only add to the confusion or provoke the dogmatic acceptance or rejection of notions that are not at all implied by the meaning of heritability.

Heredity versus Environment. Genetic and environmental factors are not properly viewed as being in opposition to each other. Nor are they an "all or none" affair. Any observable characteristic, physical or behavioral, is a phenotype, the very existence of which depends upon both genetic and environmental conditions. The legitimate question is not whether the characteristic is due to heredity *or* environment, but what proportion of the population variation in the characteristic is attributable to genotypic variation (which is H, the heritability) and what proportion is attributable to non-genetic or environmental variation in the population (which is $1-H$). For metric characteristics like stature and intelligence, H can have values between 0 and 1.

Individual versus Population. Heritability is a population statistic, describing the relative magnitude of the genetic component (or set of genetic components) in the population variance of the characteristic in question. It has no sensible meaning with reference to a measurement or characteristic in an individual. A single measurement, by definition, has no variance. There is no way of partitioning a given individual's IQ into hereditary and environmental components, as if the person inherited, say, 80 points of IQ and acquired 20 additional points from his environment. This is, of course, nonsense. *The square root of the heritability* $(\sqrt{H})$, *however, tells us the correlation between genotypes and phenotypes in the population, and this permits a probabalistic inference concerning the average amount of difference between individuals' obtained IQs and the "genotypic value" of their intelligence.* (The average correlation between phenotypes and genotypes for IQ is about .90 in European and North American Caucasian populations, as determined from summary data presented later in this paper [Table 2]. The square of this value is known as the heritability—the proportion of phenotypic variance due to genetic variation.) The principle is the same as estimating the "true" scores from obtained scores in test theory. Statements about individuals can be made only on a probabilistic basis and not with absolute certainty. Only if heritability were unity (i.e., $H = 1$) would there be a perfect correlation between obtained scores and genotypic values, in which case we could say with assurance that an individual's measured IQ perfectly represented his genotype for intelligence. This still would not mean that the phenotype could have developed without an environment, for without either heredity or environment there simply is no organism and no phenotype. Thus the statement we so often hear in discussions of individual differences—that the individual's intelligence is the product of the interaction of his heredity and his environment—is rather fatuous. It really states nothing more than the fact that the individual exists.

Constancy. From what has already been said about heritability, it must be clear that it is not a constant like π and the speed of light. H is an empirically determined population statistic, and like any statistic, its value is affected by the characteristics of the population. H will be higher in a population in which environmental variation relevant to the trait in question is small, than in a population in which there is great environmental variation. Similarly, when a population is relatively homogeneous in genetic factors but not in the environmental factors relevant to the development of the characteristic, the heritability of the characteristic in question will be lower. In short, the value of H is jointly a function of genetic and environmental variability in the population. Also, like any other statistic, it is an estimate based on a sample of the population and is therefore subject to sampling error—the smaller the sample, the greater the margin of probable error. Values of H reported in the literature do not represent what the heritability might be under any environmental conditions or in all populations or even in the same population at different times. Estimates of H are specific to the population sampled, the point in time, how the measurements were made, and the particular test used to obtain the measurements.

Measurements versus Reality. It is frequently argued that since we cannot really measure intelligence we cannot possibly determine its heritability. Whether we can or cannot measure intelligence, which is a separate issue I have already discussed, let it be emphasized that it makes no difference to the question of heritability. We do not estimate the heritability of some trait that lies hidden behind our measurements. We estimate the heritability of the phenotypes and these are the measurements themselves. Regardless of what it is that our tests measure, the heritability tells us how much of the variance in these measurements is due to genetic factors. If the test scores get at nothing genetic, the result will simply be that estimates of their heritability will not differ significantly from zero. The fact that heritability estimates based on IQs differ very significantly from zero is proof that genetic factors play a part in individual differences in IQ. To the extent that a test is not "culture-free" or "culture-fair," it will result in a lower heritability measurement. It makes no more sense to say that intelligence tests do not really measure intelligence but only *developed* intelligence than to say that scales do not really measure a person's weight but only the weight he has acquired by eating. An "environment-free" test of intelligence makes as much sense as a "nutrition-free" scale for weight.

Know All versus Know Nothing. This expression describes another confused notion: the idea that unless we can know absolutely *everything* about the genetics of intelligence we can know nothing! Proponents of this view demand that we be able to spell out in detail every single link in the chain of causality from genes (or DNA molecules) to test scores if we are to say anything about the heritability of intelligence. Determining the heritability of a characteristic does not at all depend upon a knowledge of its physical, biochemical, or physiological basis or of the precise mechanisms through which the characteristic is modified by the environment. Knowledge of these factors is, of course, important in its own right, but we need not have such knowledge to establish the genetic basis of the characteristic. Selective breeding was practiced fruitfully for centuries before anything at all was known of chromosomes and genes, and the science of quantitative genetics upon which the estimation of heritability depends has proven its value independently of advances in biochemical and physiological genetics.

Acquired versus Inherited. How can a socially defined attribute such as intelligence be said to be inherited? Or something that is so obviously acquired from the social environment as vocabulary? Strictly speaking, of course, only genes are inherited. But the brain mechanisms which are involved in learning are gene-

44

tically conditioned just as are other structures and functions of the organism. What the organism is capable of learning from the environment and its rate of learning thus have a biological basis. Individuals differ markedly in the amount, rate, and kinds of learning they evince even given equal opportunities. Consider the differences that show up when a Mozart and the average run of children are given music lessons! If a test of vocabulary shows high heritability, it only means that persons in the population have had fairly equal opportunity for learning all the words in the test, and the differences in their scores are due mostly to differences in capacity for learning. If members of the population had had very unequal exposures to the words in the vocabulary test, the heritability of the scores would be very low.

Immutability. High heritability by itself does not necessarily imply that the characteristic is immutable. Under greatly changed environmental conditions, the heritability may have some other value, or it may remain the same while the mean of the population changes. At one time tuberculosis had a very high heritability, the reason being that the tuberculosis bacilli were extremely widespread throughout the population, so that the main factor determining whether an individual contracted tuberculosis was not the probability of exposure but the individual's inherited physical constitution. Now that tuberculosis bacilli are relatively rare, difference in exposure rather than in physical predisposition is a more important determinant of who contracts tuberculosis. In the absence of exposure, individual differences in predisposition are of no consequence.

Heritability also tells us something about the locus of control of a characteristic. The control of highly heritable characteristics is usually in the organism's internal biochemical mechanisms. Traits of low heritability are usually controlled by external environmental factors. No amount of psychotherapy, tutoring, or other psychological intervention will elicit normal performance from a child who is mentally retarded because of phenylketonuria (PKU), a recessive genetic defect of metabolism which results in brain damage. Yet a child who has inherited the genes for PKU can grow up normally if his diet is controlled to eliminate certain proteins which contain phenylalanine. Knowledge of the genetic and metabolic basis of this condition in recent years has saved many children from mental retardation.

Parent-Child Resemblance. The old maxim that "like begets like" is held up as an instance of the workings of heredity. The lack of parent-child resemblance, on the other hand, is often mistakenly interpreted as evidence that a characteristic is not highly heritable. But the principles of genetics also explain the fact that often "like begets unlike." A high degree of parent-offspring resemblance, in fact, is to be expected only in highly inbred (or homozygous) strains, as in certain highly selected breeds of dogs and laboratory strains of mice. The random segregation of the parental genes in the formation of the sex cells means that the child receives a random selection of only half of each parent's genes. This fact that parent and child have only 50 percent of their genes in common, along with the effects of dominance and epistasis, insures considerable genetic dissimilarity between parent and child as well as among siblings, who also have only 50 percent of their genes in common. The fact that one parent and a child have only 50 percent of their genes in common is reflected in the average parent-offspring correlation (r_{po}) of between .50 and .60 (depending on the degree of assortative mating for a given characteristic) which obtains for height, head circumference, fingerprint ridges, intelligence, and other highly heritable characteristics. (The correlation is also between .50 and .60 for siblings on these characteristics; sibling resemblance is generally much *higher* than this for traits of *low* heritability.) The genetic correlation between the average of both parents (called the "midparent") and a single offspring $(r_{\bar{p}o})$ is the square root of the correlation for a single parent (i.e., $r_{\bar{p}o} = \sqrt{r_{po}}$). The correlation between the average of *both* parents and the average of *all* the offspring ("midchild") that they could theoretically produce $(r_{\bar{p}\bar{o}})$ is the same value as $\bar{H}_N$, i.e., heritability in the narrow sense.[4] It is noteworthy that empirical determinations of the midparent-midchild correlation $(r_{\bar{p}\bar{o}})$ in fact closely approximate the values of H as estimated by various methods, such as comparisons of twins, siblings, and unrelated children reared together.

Empirical Findings on the Heritability of Intelligence

It is always preferable, of course, to have estimates of the proportions of variance contributed by each of the components in Equation 2 than to have merely an overall estimate of H. But to obtain reliable estimates of the separate components requires large samples of persons of different kinships, such as identical twins reared together and reared apart, fraternal twins, siblings, half-siblings, parents-children,

---

[4] Heritability in the narrow sense is an estimate of the proportion of genic variance without consideration of dominance and epistasis. This contrasts with equation (3), the definition of H, which includes estimates for these two factors. Signified as H_N, heritability in the narrow sense is conceptually defined as:

$$H_N = \frac{(V_G + V_{AM})}{V_p - V_e}$$

46

cousins, and so on. The methods of quantitative genetics by which these variance components, as well as the heritability, can be calculated from such kinship data are technical matters beyond the scope of this article, and the reader must be referred elsewhere for expositions of the methodology of quantitative genetics (Cattell, 1960; Falconer, 1960; Huntley, 1966; Kempthorne, 1957; Loehlin, in press).

The most satisfactory attempt to estimate the separate variance components is the work of Sir Cyril Burt (1955, 1958), based on large samples of many kinships drawn mostly from the school population of London. The IQ test used by Burt was an English adaptation of the Stanford-Binet. Burt's results may be regarded as representative of variance components of intelligence in populations that are similar to the population of London in their degree of genetic heterogenity and in their range of environmental variation. Table 1 shows the percentage of variance due to the various components, grouped under "genetic" and "environmental," in Burt's analysis.

TABLE 1

Analysis of Variance of Intelligence Test
Scores (Burt, 1958)

Source of Variance	Percent*	
Genetic:		
Genic (additive)	40.5	(47.9)
Assortative Mating	19.9	(17.9)
Dominance & Epistasis	16.7	(21.7)
Environmental:		
Covariance of Heredity & Environment	10.6	(1.4)
Random Environmental Effects, including		
H × E interaction (V_I)	5.9	(5.8)
Unreliability (test error)	6.4	(5.3)
Total	100.0	(100.0)

* Figures in parentheses are percentages for adjusted assessments. See text for explanation.

When Burt submitted the test scores to the children's teachers for criticism on the basis of their impressions of the child's "brightness," a number of children were identified for whom the IQ was not a fair estimate of the child's ability in the teacher's judgment. These children were retested, often on a number of tests on several occasions, and the result was an "adjusted" assessment of the child's

47

IQ. The results of the analysis of variance after these adjusted assessments were made are shown in parentheses in Table 1. Note that the component most affected by the adjustments is the covariance of heredity and environment, which is what we should expect if the test is not perfectly "culture-fair." It means that the adjusted scores reduced systematic environmental sources of variance and thereby came closer to representing the children's innate ability, or, stated more technically, the adjusted scores increased the correlation between genotype and phenotype from .88 for unadjusted scores to .93 for adjusted scores. (Corrected for test unreliability these correlations become .90 and .96, respectively. And the heritabilities (H_B) for the two sets of scores are therefore $(.90)^2 = .81$ and $(.96)^2 = .93$, respectively.)

Kinship Correlations. The basic data from which variance components and heritability coefficients are estimated are correlations among individuals of different degrees of kinship. Nearly all such kinship correlations reported in the literature are summarized in Table 2. The median values of the correlations obtained in the various studies are given here. These represent the most reliable values we have for the correlations among relatives. Most of the values are taken from the survey by Erlenmeyer-Kimling and Jarvik (1963), and I have supplemented these with certain kinship correlations not included in their survey and reported in the literature since their review (e.g., Burt, 1966, p. 150). The Erlenmeyer-Kimling and Jarvik (1963) review was based on 52 independent studies of the correlations of relatives for tested intellectual abilities, involving over 30,000 correlational pairings from 8 countries in 4 continents, obtained over a period of more than two generations. The correlations were based on a wide variety of mental tests, administered under a variety of conditions by numerous investigators with contrasting views regarding the importance of heredity. The authors conclude: "Against this pronounced heterogeneity, which should have clouded the picture, and is reflected by the wide range of correlations, a clearly definite consistency emerges from the data. The composite data are compatible with the polygenic hypothesis which is generally favored in accounting for inherited differences in mental ability" (Erlenmeyer-Kimling & Jarvik, 1963, p. 1479).

The compatibility with the polygenic hypothesis to which the authors (as outlined earlier on p. 53) refer can be appreciated in Table 2 by comparing the median values of the obtained correlations with the sets of theoretical values shown in the last two columns. The first set (Theoretical Value[1]) is based on calculations by Burt (1966), using the methods devised by Fisher for estimating

TABLE 2

Correlations for Intellectual Ability: Obtained and Theoretical Values

Correlations Between	Number of Studies	Obtained Median r*	Theoretical Value[1]	Theoretical Value[2]
Unrelated Persons				
Children reared apart	4	−.01	.00	.00
Foster parent and child	3	+.20	.00	.00
Children reared together	5	+.24	.00	.00
Collaterals				
Second Cousins	1	+.16	+ .14	+ .063
First Cousins	3	+.26	+ .18	+ .125
Uncle (or aunt) and nephew (or niece)	1	+.34	+ .31	+ .25
Siblings, reared apart	3	+.47	+ .52	+ .50
Siblings, reared together	36	+.55	+ .52	+ .50
Dizygotic twins, different sex	9	+.49	+ .50	+ .50
Dizygotic twins, same sex	11	+.56	+ .54	+ .50
Monozygotic twins, reared apart	4	+.75	+1.00	+1.00
Monozygotic twins, reared together	14	+.87	+1.00	+1.00
Direct Line				
Grandparent and grandchild	3	+.27	+ .31	+ .25
Parent (as adult) and child	13	+.50	+ .49	+ .50
Parent (as child) and child	1	+.56	+ .49	+ .50

* Correlations not corrected for attenuation (unreliability).

[1] Assuming assortative mating and partial dominance.

[2] Assuming random mating and only additive genes, i.e., the simplest possible polygenic model.

kinship correlations for physical characteristics involving assortative mating and some degree of dominance. The second set (Theoretical Value[2]) of theoretical values is based on the simplest possible polygenic model, assuming random mating and nothing but additive gene effects. So these are the values one would expect if genetic factors alone were operating and the trait variance reflected no environmental influences whatsoever.

First of all, one can note certain systematic departures of the obtained correlations from the theoretical values. These departures are presumably due to nongenetic or environmental influences. The orderly nature of these environmental effects, as reflected in the Erlenmeyer-Kimling and Jarvik median correlations, can be highlighted by graphical presentation, as shown in Figure 6. Note that the condition of being reared together or reared apart has the same effect on the difference in magnitudes of the correlations for the various kinships. (The

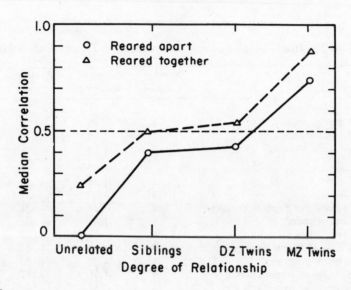

FIGURE 6.

Median values of all correlations reported in the literature up to 1963 for the indicated kinships. (After Erlenmeyer-Kimling & Jarvik, 1963.) Note consistency of difference in correlations for relatives reared together and reared apart.

slightly greater difference for unrelated children is probably due to the fact of selective placement by adoption agencies, that is, the attempt to match the child's intelligence with that of the adopting parents.)

Heritability Estimates. By making certain comparisons among the correlations shown in Table 2 and Figure 6, one can get some insight into how heritability is estimated. For example, we see that the correlation between identical or monozygotic (MZ) twins reared apart is .75. Since MZ twins develop from a single fertilized ovum and thus have exactly the same genes, any difference between the twins must be due to nongenetic factors. And if they are reared apart in uncorrelated environments, the difference between a perfect correlation (1.00) and the obtained correlation (.75) gives an estimate of the proportion of the variance in IQs attributable to environmental differences: $1.00 - 0.75 = 0.25$. Thus 75 percent of the variance can be said to be due to genetic variation (this is the heritability) and 25 percent to environmental variation. Now let us go to the other extreme and look at unrelated children reared together. They have no genetic inheritance in common, but they are reared in a common environment. Therefore the correlation between such children will reflect the environment. As seen in Table 2, this correlation is 0.24. Thus, the proportion of IQ variance due to environment is .24; and the remainder, 1.00 — .24 = .76 is due to heredity. There is quite good agreement between the two estimates of heritability.

Another interesting comparison is between MZ twins reared together ($r = .87$) and reared apart ($r = .75$). If 1.00 — .75 = .25 (from MZ twins reared apart) estimates the total environmental variance, then 1.00 — .87 = .13 (from MZ twins reared together) is an estimate of the environmental variance *within families* in which children are reared together. Thus the difference between .25 — .13 = .12 is an estimate of the environmental variance *between families*.

The situation is relatively simple when we deal only with MZ twins, who are genetically identical, or with unrelated children, who have nothing in common genetically. But in order to estimate heritability from any of the other kinship correlations, much more complex formulas are needed which would require much more explanation than is possible in this article. I have presented elsewhere a generalized formula for estimating heritability from any two kinship correlations where one kinship is of a higher degree than the other (Jensen, 1967a). I applied this heritability formula to all the correlations for monozygotic and dizygotic (half their genes in common) twins reported in the literature and found an average heritability of .80 for intelligence test scores. (The correlations from which this heritability estimate was derived were corrected for unreliability.) Environmental differences *between* families account for .12 of the total variance, and differences *within* families account for .08. It is possible to derive an overall heritability coefficient from all the kinship correlations given in Table 2. This composite value of H is .77, which becomes .81 after correction for unreliability (assuming an average test reliability of .95). This represents probably the best single overall estimate of the heritability of measured intelligence that we can make. But, as pointed out previously, this is an average value of H about which there is some dispersion of values, depending on such variables as the particular tests used, the population sampled, and sampling error.

Identical Twins Reared Apart. The conceptually simplest estimate of heritability is, of course, the correlation between identical twins reared apart, since, if their environments are uncorrelated, all they have in common are their genes. The correlation (corrected for unreliability) in this case is the same as the heritability as defined in Equation 3. There have been only three major studies of MZ twins separated early in life and reared apart. All three used individually administered intelligence tests. The correlation between Stanford-Binet IQs of 19 pairs of MZ twins reared apart in a study by Newman, Freeman, and Holzinger (1937) was .77 (.81 corrected for unreliability). The correlation between 44 pairs of MZ twins reared apart on a composite score based on a vocabulary test and Raven's Progressive Matrices was .77 (.81 corrected) in a study by Shields (1962). The correlation between 53 pairs on the Stanford-Binet was .86 (.91 corrected) in a study by Burt (1966). Twin correlations in the same group for height and for weight were .94 and .88, respectively.

The Burt study is perhaps the most interesting, for four reasons: (a) it is based on the largest sample; (b) the IQ distribution of the sample had a mean of 97.8 and a standard deviation of 15.3—values very close to those of the general population; (c) all the twin pairs were separated at birth or within their first six months of life; and (d) most important, the separated twins were spread over the entire range of socioeconomic levels (based on classification in terms of the six socioeconomic categories of the English census), and there was a slight, though nonsignificant, negative correlation between the environmental ratings of the separated twin pairs. When the twin pairs were rated for differences in the cultural conditions of their rearing, these differences correlated .26 with the differences in their IQs. Differences between the material conditions of their homes correlated .16 with IQ differences. (The corresponding correlations for a measure of scholastic attainments were .74 and .37, respectively. The correlation between the twins in scholastic attainments was only .62, indicating a much lower heritability than for IQ.)

Foster Parents versus Natural Parents. Children separated from their true parents shortly after birth and reared in adoptive homes show almost the same degree of correlation with the intelligence of their biological parents as do children who are reared by their own parents. The correlations of children with their foster parents' intelligence range between 0 and .20 and are seldom higher than this even when the adoption agency attempts selective placement (e.g., Honzik, 1957). Parent-child correlations gradually increase from zero at 18 months of age to an asymptotic value close to .50 between ages 5 and 6 (Jones, 1954), and this is true whether the child is reared by his parents or not.

Direct Measurement of the Environment. Another method for getting at the relative contribution of environmental factors to IQ variance is simply by correlating children's IQs with ratings of their environment. This can be legitimately done only in the case of adopted children and where there is evidence that selective placement by the adoption agencies is negligible. Without these conditions, of course, some of the correlation between the children and their environmental ratings will be due to genetic factors. There are two large-scale studies in the literature which meet these criteria. Also, both studies involved adopting parents who were representative of a broad cross section of the U.S. Caucasian population with respect to education, occupation, and socioeconomic level. It is probably safe to say that not more than five percent of the U.S. Caucasian population falls outside the range of environmental variation represented in the samples in these two studies. The study by Leahy (1935) found an average correlation of .20 between the IQs of adopted children and a number of indices of the "goodness" of their environment, including the IQs and education of both adopting parents, their socioeconomic status, and the cultural amenities in the home. Leahy concluded from this that the environmental ratings accounted for 4 percent (i.e., the square of $r = .20$) of the variance in the adopted children's Stanford-Binet IQs, and that 96 percent of the variance remained to be accounted for by other factors. The main criticisms we can make of this study are, first, that the environmental indices were not sufficiently "fine-grained" to register the subtleties of environmental variation and of the qualities of parent-child relationship that influence intellectual development, and, second, that the study did not make use of the technique of multiple correlation, which would show the total contribution to the variance of all the separate environmental indices simultaneously. A multiple correlation is usually considerably greater than merely the average of all the correlations for the single variables.

A study by Burks (1928) meets both these objections. To the best of my knowledge no study before or since has rated environments in any more detailed and fine-grained manner than did Burks'. Each adoptive home was given 4 to 8 hours of individual investigation. As in Leahy's study, Burks included intelligence measures on the adopting parents as part of the children's environment, an environment which also included such factors as the amount of time the parents spent helping the children with their school work, the amount of time spent reading to the children, and so on. The multiple correlation (corrected for unreliability) between Burks' various environmental ratings and the adopted children's Stanford-Binet IQs was .42. The square of this correlation is .18, which represents the proportion of IQ variance accounted for by Burks' environmental measurements. This value comes very close to the environmental variance estimated in direct heritability analyses based on kinship correlations.

Burks translated her findings into the conclusion that the total effect of environmental factors one standard deviation up or down the environmental scale is only about 6 IQ points. This is an interesting figure, since it is exactly half the 12 point IQ difference found on the average between normal siblings reared together by their own parents. Siblings differ genetically, of course, having only about half their genes in common. If all the siblings in every family were divided into two groups—those above and those below the family average—the IQ distributions of the two groups would appear as shown in Figure 7. Though the average difference is only 12 IQ points, note the implications in the proportions of

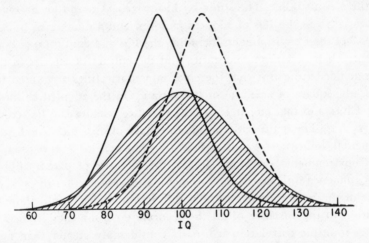

FIGURE 7.

IQ distributions of siblings who are below (solid curve) or above (dashed curve) their family average. The shaded curve is the IQ distribution of randomly selected children.

each group falling into the upper and lower ranges of the IQ scale. It would be most instructive to study the educational and occupational attainments of these two groups, since presumably they should have about the same environmental advantages.

Another part of Burks' study consisted of a perfectly matched control group of parents rearing their own children, for whom parent-child correlations were obtained. Sewall Wright (1931) performed a heritability analysis on these parent-child and IQ-environment correlations and obtained a heritability coefficient of .81.

Effects of Inbreeding on Intelligence

One of the most impressive lines of evidence for the involvement of genetic factors in intelligence comes from study of the effects of inbreeding, that is, the mating of relatives. In the case of polygenic characteristics the direction of the effect of inbreeding is predictable from purely genetic considerations. All individuals carry in their chromosomes a number of mutant or defective genes. These genes are almost always recessive, so they have no effect on the phenotype unless by rare chance they match up with another mutant gene at the same locus on a homologous chromosome; in other words, the recessive mutant gene at a given locus must be inherited from both the father and mother in order to affect the phenotype. Since such mutants are usually defective, they do not enhance the phenotypic expression of the characteristic but usually degrade it. And for polygenic characteristics we would expect such mutants to lower the metric value of the characteristics by graded amounts, depending upon the number of paired mutant recessives. If the parents are genetically related, there is a greatly increased probability that the mutant recessives at given loci will be paired in the offspring. The situation is illustrated in Figure 8, which depicts in a simplified way a pair of homologous chromosomes inherited by an individual from a mother (M) and father (F) who are related (Pair A) and a pair of chromosomes inherited from unrelated parents (Pair B). The blackened spaces represent recessive genes. Although both pairs contain equal numbers of recessives, more of them are at the same loci in Pair A than in Pair B. Only the paired genes degrade the characteristics' phenotypic value.

A most valuable study of this genetic phenomenon with respect to intelligence was carried out in Japan after World War II by Schull and Neel (1965). The study illustrates how strictly sociological factors, such as mate selection, can have extremely important genetic consequences. In Japan approximately five percent of all marriages are between cousins. Schull and Neel studied the offspring of marriages of first cousins, first cousins once removed, and second cousins. The parents were statistically matched with a control group of unrelated parents for age and socioeconomic factors. Children from the cousin marriages and the control children from unrelated parents (total $N = 2,111$) were given the Japanese version of the Wechsler Intelligence Scale for Children (WISC). The degree of consanguinity represented by the cousin marriages in this study had the effect of depressing WISC IQs by an average of 7.4 percent, making the mean of the inbred group nearly 8 IQ points lower than the mean of the control group. Assuming normal distributions of IQ, the effect is shown in

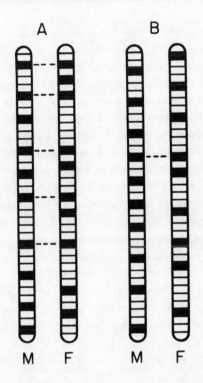

FIGURE 8.

Simplified schema of chromosomes, illustrating the pairing of recessive (mutant) genes (black spaces) in homologous chromosomes from mother (M) and father (F). Pair A has five pairs of recessives in the same loci on the chromosome, Pair B has only one such pair.

Figure 9, and illustrates the point that the most drastic consequences of group mean differences are to be seen in the tails of the distributions. In the same study a similar depressing effect was found for other polygenic characteristics such as several anthropometric and dental variables.

The mating of relatives closer than cousins can produce a markedly greater reduction in offspring's IQs. Lindzey (1967) has reported that almost half of a group of children born to so-called nuclear incest matings (brother-sister or father-daughter) could not be placed for adoption because of mental retardation and other severe defects which had a relatively low incidence among the offspring of unrelated parents who were matched with the incestuous parents in

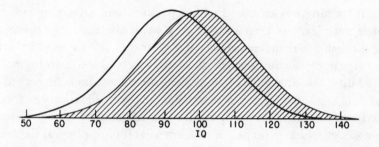

FIGURE 9.

The average effect of inbreeding to the degree of 1st, 1 1/2, and 2nd cousin matings on the IQ distribution of offspring (heavy line). Shaded curve is the IQ distribution of the offspring of nonconsanguinous matings. (After Schull & Neel, 1965.)

intelligence, socioeconomic status, age, weight, and stature. In any geographically confined population where social or legal regulations on mating are lax, where individuals' paternity is often dubious, and where the proportion of half-siblings within the same age groups is high, we would expect more inadvertent inbreeding, with its unfavorable genetic consequences, than in a population in which these conditions exist to a lesser degree.

Heritability of Special Mental Abilities. When the general factor, or *g*, is removed from a variety of mental tests, the remaining variance is attributable to a number of so-called "group factors" or "special abilities." The tests of special abilities that have been studied most thoroughly with respect to their heritability are Thurstone's Primary Mental Abilities: Verbal, Space, Number, Word Fluency, Memory, and Perceptual Speed. Vandenberg (1967) has reviewed the heritability studies of these tests and reports that the *H* values range from near zero to about .75, with most values of *H* between .50 and .70. Vandenberg devised a method for estimating the genetic components of these special abilities which are completely independent of *g*. He concluded that at least four of the Primary Mental Abilities (Number, Verbal, Space, and Word Fluency) independently have significant hereditary components.

There have been few studies of the heritability of noncognitive skills, but a study by McNemar (see Bilodeau, 1966, Ch. 3) of motor skill learning indicates that heritabilities in this sphere may be even higher than for intelligence. The motor skill learning was measured with a pursuit-rotor, a tracking task in which the subject must learn to keep a stylus on a metal disc about the size of a nickel rotating through a circumference of about 36 inches at 60 rpm. The percentage of time "on target" during the course of practice yields a learning measure of high reliability, showing marked individual differences both in rate of acquisition and final asymptote of this perceptual-motor skill. Identical twins correlated .95 and fraternal twins .51 on pursuit-rotor learning, yielding a heritability coefficient of .88, which is very close to the heritability of physical stature.

Heritability of Scholastic Achievement. The heritability of measures of scholastic achievement is much less, on the average, than the heritability of intelligence. In reviewing all the twin studies in the literature containing relevant data, I concluded that individual differences in scholastic performance are determined less than half as much by heredity as are differences in intelligence (Jensen, 1967a).[5] The analysis of all the twin studies on a variety of scholastic measures gives an average *H* of .40. The environmental variance of 60 percent can be partitioned into variance due to environmental differences *between* families, which is 54 percent, and differences *within* families of 6 percent. But it should also be noted that the heritability estimates for scholastic achievement vary over a much wider range than do *H* values for intelligence. In general, *H* for scholastic achievement increases as we go from the primary grades up to high school and it is somewhat lower for relatively simple forms of learning (e.g., spelling and arithmetic compu-

[5] After this article went to press I received a personal communication from Professor Lloyd G. Humphreys who pointed out some arguments that indicate I may have underestimated the heritability of scholastic achievement and that its heritability may actually be considerably closer to the heritability of intelligence. The argument involves two main points: (1) the fact that some of the achievement tests that entered into the average estimate of heritability are tests of specific achievements, rather than omnibus achievement tests, and therefore would correspond more to the separate subscales of the usual intelligence tests, which are known to have somewhat lower heritabilities than the composite scores; and (2) scores on some of the achievement tests are age-related, so that fraternal twin correlations, in relation to other kinship correlations, are unduly inflated by common factor of age. When age is partialled out of the MZ and DZ twin correlations, the estimate of heritability based on MZ and DZ twin comparisons is increased. However, an omnibus achievement test (Stanford Achievement) yielding an overall Educational Age score had a heritability of only .46 (as compared with .63 for Stanford-Binet IQ and .70 for Otis IQ based on the same set of MZ and DZ twins), with age partialled out of the twin correlations (Newman, Freeman, and Holzinger, 1937, p. 97). Rank in high school graduating class, which is an overall index of scholastic performance and is little affected by age yields heritability coefficients below .40 in a nationwide sample (Nichols & Bilbro, 1966). The issue clearly needs further study, but the best conclusion that can be drawn from the existing evidence, I believe, still is that the heritability of scholastic achievement is less than for intelligence, but the amount of the difference cannot be precisely estimated at present.

tation) than for more complex learning (e.g., reading comprehension and arithmetic problem solving). Yet large-sample twin data from the National Merit Scholarship Corporation show that the *between families* environmental component accounts for about 6o percent of the variance in students' rank in their high school graduating class. This must mean that there are strong family influences which cause children to conform to some academic standard set by the family and which reduce variance in scholastic performance among siblings reared in the same family. Unrelated children reared together are also much more alike in school performance than in intelligence. The common finding of a negative correlation between children's IQ and the amount of time parents report spending in helping their children with school work is further evidence that considerable family pressures are exerted to equalize the scholastic performance of siblings. This pressure to conform to a family standard shows up most conspicuously in the small *within families* environmental variance component on those school subjects which are most susceptible to improvement by extra coaching, such as spelling and arithmetic computation.

The fact that scholastic achievement is considerably less heritable than intelligence also means that many other traits, habits, attitudes, and values enter into a child's performance in school besides just his intelligence, and these non-cognitive factors are largely environmentally determined, mainly through influences within the child's family. This means there is potentially much more we can do to improve school performance through environmental means than we can do to change intelligence per se. Thus it seems likely that if compensatory education programs are to have a beneficial effect on achievement, it will be through their influence on motivation, values, and other environmentally conditioned habits that play an important part in scholastic performance, rather than through any marked direct influence on intelligence per se. The proper evaluation of such programs should therefore be sought in their effects on actual scholastic performance rather than in how much they raise the child's IQ.

How the Environment Works

Environment as a Threshold

All the reports I have found of especially large upward shifts in IQ which are explicitly associated with environmental factors have involved young children, usually under six years of age, whose initial social environment was deplorable to a greater extreme than can be found among any children who are free to interact with other persons or to run about out-of-doors. There can be no doubt that moving children from an extremely deprived environment to good average environmental circumstances can boost the IQ some 20 to 30 points and in certain extreme rare cases as much as 60 or 70 points. On the other hand, children reared in rather average circumstances do not show an appreciable IQ gain as a result of being placed in a more culturally enriched environment. While there are reports of groups of children going from below average up to average IQs as a result of environmental enrichment, I have found no report of a group of children being given permanently superior IQs by means of environmental manipulations. In brief, it is doubtful that psychologists have found consistent evidence for any social environmental influences short of extreme environmental isolation which have a marked systematic effect on intelligence. This suggests that the influence of the quality of the environment on intellectual development is not a linear function. Below a certain threshold of environmental adequacy, deprivation can have a markedly depressing effect on intelligence. But above this threshold, environmental variations cause relatively small differences in intelligence. The fact that the vast majority of the populations sampled in studies of the heritability of intelligence are above this threshold level of environmental adequacy accounts for the high values of the heritability estimates and the relatively small proportion of IQ variance attributable to environmental influences.

The environment with respect to intelligence is thus analogous to nutrition with respect to stature. If there are great nutritional lacks, growth is stunted, but above a certain level of nutritional adequacy, including minimal daily requirements of minerals, vitamins, and proteins, even great variations in eating habits will have negligible effects on persons' stature, and under such conditions most of the differences in stature among individuals will be due to heredity.

When I speak of subthreshold environmental deprivation, I do not refer to a mere lack of middle-class amenities. I refer to the extreme sensory and motor restrictions in environments such as those described by Skeels and Dye (1939) and Davis (1947), in which the subjects had little sensory stimulation of any kind and little contact with adults. These cases of extreme social isolation early in life showed great deficiencies in IQ. But removal from social deprivation to a good, average social environment resulted in large gains in IQ. The Skeels and Dye orphanage children gained in IQ from an average of 64 at 19 months of age to 96 at age 6 as a result of being given social stimulation and placement in good homes between 2 and 3 years of age. When these children were followed up as adults, they were found to be average citizens in their communities, and their own children had an average IQ of 105 and were doing satisfactorily in school. A far more extreme case was that of Isabel, a child who was confined and reared in an attic up to the age of six by a deaf-mute mother, and who had an IQ of about 30 at age 6. When Isabel was put into a good environment at that age, her IQ became normal by age 8 and she was able to perform as an average student throughout school (Davis, 1947). Extreme environmental deprivation thus need not permanently result in below average intelligence.

These observations are consistent with studies of the effects of extreme sensory deprivation on primates. Monkeys raised from birth under conditions of total social isolation, for example, show no indication when compared with normally raised controls, of any permanent impairment of ability for complex discrimination learning, delayed response learning, or learning set formation, although the isolated monkeys show severe social impairment in their relationships to normally reared monkeys (Harlow & Griffin, 1965).

Thoughtful scrutiny of all these studies of extreme environmental deprivation leads to two observations which are rarely made by psychologists who cite the studies as illustrative explanations of the low IQs and poor scholastic performance of the many children called culturally disadvantaged. In the first place, typical culturally disadvantaged children are not reared in anything like the degree of sensory and motor deprivation that characterizes, say, the children of the Skeels study. Secondly, the IQs of severely deprived children are markedly depressed even at a very early age, and when they are later exposed to normal environmental stimulation, their IQs rise rapidly, markedly, and permanently. Children called culturally disadvantaged, on the other hand, generally show no early deficit and are usually average and sometimes precocious on perceptual-motor tests administered before two years of age. The orphanage children described in Skeels' study are in striking contrast to typical culturally disadvantaged children of the same age. Also, culturally disadvantaged children usually show a slight initial gain in IQ after their first few months of exposure to the environmental enrichment afforded by school attendance, but, unlike Skeels' orphans, they soon lose this gain, and in a sizeable proportion of children the initial IQ gain is followed by a gradual decline in IQ throughout the subsequent years of schooling. We do not know how much of this decline is related to environmental or hereditary factors. We do know that with increasing age children's IQs increasingly resemble their parents' rank order in intelligence whether they are reared by them or not, and therefore with increasing age we should expect greater and more reliable differentiation among children's IQs as they gravitate toward their genotypic values

(Honzik, 1957). Of course, the gravitating effect is compounded by the fact that less intelligent parents are also less apt to provide the environmental conditions conducive to intellectual development in the important period between ages 3 and 7, during which children normally gain increasing verbal control over their environment and their own behavior. (I have described some of these environmental factors in detail elsewhere [Jensen, 1968e].)

Heber, Dever, and Conry (1968) have obtained data which illustrate this phenomenon of children's gravitation toward the parental IQ with increasing age. They studied the families of 88 low economic class Negro mothers residing in Milwaukee in a set of contiguous slum census tracts, an area which yields the highest known prevalence of identified retardation in the city's schools. Although these tracts contribute about 5 percent of the schools' population, they account for about one-third of the school children classed as mentally retarded (IQ below 75). The sample of 88 mothers was selected by taking 88 consecutive births in these tracts where the mother already had at least one child of age six. The 88 mothers had a total of 586 children, excluding their newborns. The percentage of mothers with IQs of 80 or above was 54.6; 45.4 percent were below IQ 80. The IQs of the children of these two groups of mothers were plotted as a function of the children's age. The results are shown in Figure 10. Note that only the children whose mothers' IQs are below 80 show a systematic decline in IQ as well as a short-lived spurt of several points at the age of entrance into school. At six years of age and older, 80.8 percent of the children with IQs below 80 were those whose mothers had IQs below 80.

It is far from certain or even likely that all such decline in IQ is due to environmental influences rather than to genetic factors involved in the growth rate of intelligence. Consistent with this interpretation is the fact that the heritability of intelligence measures increases with age. We should expect just the opposite if environmental factors alone were responsible for the increasing IQ deficit of markedly below average groups. A study by Wheeler (1942) suggests that although IQ may be raised at all age levels by improving the environment, such improvements do not counteract the decline in the IQ of certain below-average groups. In 1940 Wheeler tested over 3000 Tennessee mountain children between the ages of 6 and 16 and compared their IQs with children in the same age range who had been given the same tests in 1930, when the average IQ and standard of living in this area would characterize the majority of the inhabitants as "culturally deprived." During the intervening 10 years state and federal intervention in this area brought about great improvements in economic conditions, standards

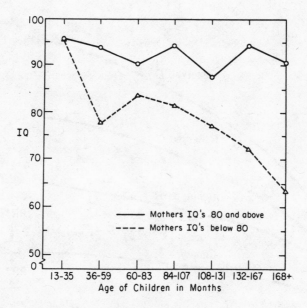

FIGURE 10.

Mean IQs of 586 children of 88 mothers as a function of age of children. (Heber, Dever, & Conry, 1968.)

of health care, and educational and cultural opportunities, and during the same period the average IQ for the region increased 10 points, from 82 to 92. But the decline in IQ from age 6 to age 16 was about the same in 1940 (from 103 to 80) as in 1930 (from 95 to 74).

Reaction Range. Geneticists refer to the concept of reaction range (RR) in discussing the fact that similar genotypes may result in quite different phenotypes depending on the favorableness of the environment for the development of the characteristic in question. Of further interest to geneticists is the fact that different genotypes may have quite different reaction ranges; some genotypes may be much more buffered against environmental influences than others. Different genetic strains can be unequal in their susceptibility to the same range of environmental variation, and when this is the case, the strains will show dissimilar heritabilities on the trait in question, the dissimilarity being accentuated by increasing environmental variation. Both of these aspects of the reaction range concept are illustrated hypothetically with respect to IQ in Figure 11.

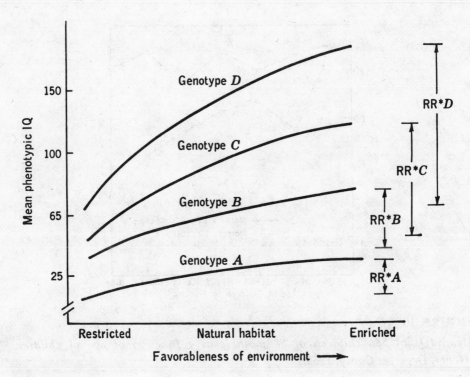

FIGURE 11.

Scheme of the reaction range concept for four hypothetical genotypes. RR denotes the presumed reaction range for phenotypic IQ. Note: Large deviations from the "natural habitat" have a low probability of occurrence. (From Gottesman, 1963.)

The above discussion should serve to counter a common misunderstanding about quantitative estimates of heritability. It is sometimes forgotten that such estimates actually represent *average* values in the population that has been sampled and they do not necessarily apply either to differences *within* various subpopulations or to differences *between* subpopulations. In a population in which an overall H estimate is, say, .80, we may find a certain group for which H is only .70 and another group for which H is .90. All the major heritability studies reported in the literature are based on samples of white European and North American populations, and our knowledge of the heritability of intelligence in different racial and cultural groups within these populations is nil. For example, no adequate heritability studies have been based on samples of the Negro population of the United States. Since some genetic strains may be more buffered from environmental influences than others, it is not sufficient merely to equate the environments of various subgroups in the population to infer equal heritability of some characteristic in all of them. The question of whether heritability estimates can contribute anything to our understanding of the relative importance of genetic and environmental factors in accounting for average phenotypic differences between racial groups (or any other socially identifiable groups) is too complex to be considered here. I have discussed this problem in detail elsewhere and concluded that heritability estimates could be of value in testing certain specific hypotheses in this area of inquiry, provided certain conditions were met and certain other crucial items of information were also available (Jensen, 1968c).

Before continuing discussion of environmental factors we must guard against one other misunderstanding about heritability that sometimes creeps in at this point. This is the notion that because so many different environmental factors and all their interactions influence the development of intelligence, by the time the child is old enough to be tested, these influences must totally bury or obscure all traces of genetic factors—the genotype must lie hidden and inaccessible under the heavy overlay of environmental influences. If this were so, of course, the obtained values of H would be very close to zero. But the fact that values of H for intelligence are usually quite high (in the region of .70 to .90) means that current intelligence tests can, so to speak, "read through" the environmental "overlay."

Physical versus Social Environment

The value $1 - H$, which for IQ generally amounts to about .20, can be called E, the proportion of variance due to nongenetic factors. There has been a pronounced tendency to think of E as being wholly associated with individuals' social and interpersonal environment, child rearing practices, and differences in educational and cultural opportunities afforded by socioeconomic status. It is certain, however, that these sociological factors are not responsible for the whole of E and it is not improbable that they contribute only a minor portion of the E variance in the bulk of our population. Certain physical and biological environmental factors may be at least as important as the social factors in determining individual differences in intelligence. If this is true, advances in medicine, nutrition, prenatal care, and obstetrics may contribute as much or more to improving intelligence as will manipulation of the social environment.

Prenatal Environment of Twins. A little known fact about twins is that they average some 4 to 7 points lower in IQ than singletons (Vandenberg, 1968). The difference also shows up in scholastic achievement, as shown in the distribution of reading scores of twin and singleton girls in Sweden (Figure 12).

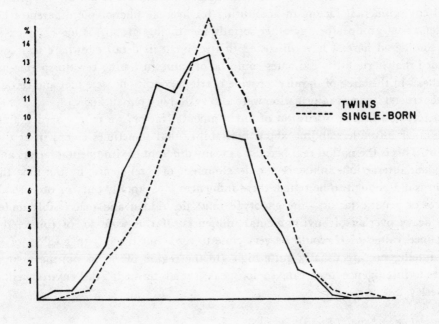

FIGURE 12.

Distribution of reading scores of twins and single children (all girls). (Husén, 1960.)

If this phenomenon were due entirely to differences between twins and singletons in the amount of individual attention they receive from their parents, one might expect the twin-singleton difference to be related to the family's socioeconomic status. But there seems to be no systematic relationship of this kind. The largest study of the question, summarized in Figure 13, shows about the same average amount of twin-singleton IQ disparity over a wide range of socioeconomic groups.

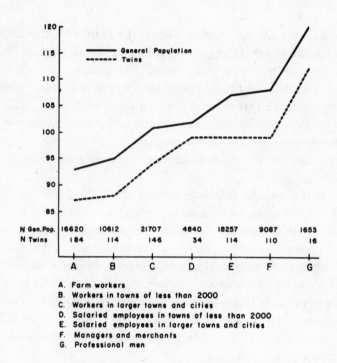

FIGURE 13.

Distribution of IQs by occupation of father, for twins and singletons. (Zazzo, 1960.)

Three other lines of evidence place the locus of this effect in the prenatal environment. Monozygotic twins are slightly lower in IQ than dizygotic twins (Stott, 1960, p. 98), a fact which is consistent with the finding that MZ twins have a higher mortality rate and greater disparity in birth weights than DZ twins, suggesting that MZ twins enjoy less equal and less optimal intrauterine conditions than DZ twins or singletons. Inequalities in both intrauterine space and fetal nutrition probably account for this. Also, boy twins are significantly lower in IQ than girl twins, which conforms to the well known greater vulnerability of male infants to prenatal impairment (Stott, 1960). Finally, the birth weight of infants, when matched for gestational age, is slightly but significantly correlated with later IQ, and the effect is independent of sociocultural factors (Churchill, Neff,

& Caldwell, 1966). In pairs of identical twins, the twin with the lower birth weight usually has the lower IQ (by 5 to 7 points on the average) at school age. This is true both in white and in Negro twins. The birth-weight differences are reflected in all 11 subtests of the Wechsler Intelligence Scale for Children and are slightly greater on the Performance than on the Verbal tests (Willerman & Churchill, 1967). The investigators interpret these findings as suggesting that nutrient supplies may be inadequate for proper body and brain development in twin pregnancies, and that the unequal sharing of nutrients and space stunts one twin more than its mate.

Thus, much of the average difference between MZ twins, whether reared together or reared apart, seems to be due to prenatal environmental factors. The real importance of these findings, of course, lies in their implications for the possible role of prenatal environment in the development of all children. It is not unlikely that there are individual maternal differences in the adequacy of the prenatal environment. If intrauterine conditions can cause several points of IQ difference between twins, it is not hard to imagine that individual differences in prenatal environments could also cause IQ differences in single born children and might therefore account for a substantial proportion of the total environmental variance in IQ.

Abdominal Decompression. There is now evidence that certain manipulations of the intrauterine environment can affect the infant's behavioral development for many months after birth. A technique known as abdominal decompression was invented by a professor of obstetrics (Heyns, 1963), originally for the purpose of making women experience less discomfort in the latter months of their pregnancy and also to facilitate labor and delivery. For about an hour a day during the last three or four months of pregnancy, the woman is placed in a device that creates a partial vacuum around her abdomen, which greatly reduces the intrauterine pressure. The device is used during labor up to the moment of delivery. Heyns has applied this device to more than 400 women. Their infants, as compared with control groups who have not received this treatment, show more rapid development in their first two years and manifest an overall superiority in tests of perceptual-motor development. They sit up earlier, walk earlier, talk earlier, and appear generally more precocious than their own siblings or other children whose mothers were not so treated. At two years of age the children in Heyns' experiment had DQs (developmental quotients) some 30 points higher than the control children (in the general population the mean DQ is 100, with a standard deviation of 15). Heyns explains the effects of maternal abdominal decompression on the child's early development in terms of the reduction of intrauterine pressure, which results in a more optimal blood supply to the fetus and also lessens the chances of brain damage during labor. (The intrauterine pressure on the infant's head is reduced from about 22 pounds to 8 pounds.) Results on children's later IQs have not been published, but correspondence with Professor Heyns and verbal reports from visitors to his laboratory inform me that there is no evidence that the IQ of these children is appreciably higher beyond age 6 than that of control groups. If this observation is confirmed by the proper methods, it should not be too surprising in view of the negligible correlations normally found between DQs and later IQs. But since abdominal decompression results in infant precocity, one may wonder to what extent differences in intrauterine pressure are responsible for normal individual and group differences in infant precocity. Negro infants, for example, are more precocious in development (as measured on the Bayley Scales) in their first year or two than Caucasian infants (Bayley, 1965a). Infant precocity would seem to be associated with more optimal intrauterine and perinatal conditions. This conjecture is consistent with the finding that infants whose prenatal and perinatal histories would make them suspect of some degree of brain damage show lower DQs on the Bayley Scales than normal infants (Honzik, 1962). Writers who place great emphasis on the hypothesis of inadequate prenatal care and complications of pregnancy to account for the lower average IQ of Negroes (e.g., Bronfenbrenner, 1967) are also obliged to explain why these unfavorable factors do not also depress the DQ below average in Negro infants, as do such factors as brain damage and prenatal and infant malnutrition (Cravioto, 1966). Since all such environmental factors should lower the heritability of intelligence in any segment of the population in which they are hypothesized to play an especially significant role, one way to test the hypothesis would be to compare the heritability of intelligence in that segment of the population for which extra environmental factors are hypothesized with the heritability in other groups for whom environmental factors are supposedly less accountable for IQ variance.

A Continuum of Reproductive Casualty. A host of conditions associated with reproduction which are known to differ greatly across socioeconomic levels have been hypothesized as causal factors in average intellectual differences. There is no doubt about the fact of the greater prevalence in poverty areas of conditions unfavorable to optimal pregnancy and safe delivery. The question that remains unanswered is the amount of IQ variance associated with these conditions predisposing to reproductive casualty. The disadvantageous factors most highly associated with social conditions are: pregnancies at early ages, teenage deliveries, pregnancies in close succession, a large number of pregnancies, and pregnancies that occur late in the woman's reproductive life (Graves, Freeman, & Thompson, 1968). These conditions are related to low birth weight, prematurity, increased infant mortality, prolonged labor, toxemia, anemia, malformations, and mental deficiency in the offspring. Since all of these factors have a higher incidence in low socioeconomic groups and in certain ethnic groups (Negroes, American Indians, and Mexican-Americans) in the United States, they probably account for some proportion of the group differences in IQ and scholastic performance, but just how much of the true differences they may account for no one really knows at present. It is interesting that Jewish immigrants, whose offspring are usually found to have a higher mean IQ than the general population, show fewer disadvantageous reproductive conditions and have the lowest infant mortality rates of all ethnic groups, even when matched with other immigrant and native born groups on general environmental conditions (Graves et al., 1968).

Although disadvantageous reproductive factors occur differentially in different segments of the population, it is not at all certain how much they are responsible for the IQ differences between social classes and races. It is reported by the National Institute of Neurological Diseases and Blindness, for example, that when all cases of mental retardation that can be reasonably explained in terms of known complications of pregnancy and delivery, brain damage, or major gene and chromosomal defects are accounted for, there still remain 75 to 80 percent of the cases who show no such specific causes and presumably represent just the lower end of the normal polygenic distribution of intelligence (Research Profile No. 11, 1965). Buck (1968) has argued that it still remains to be proven that a degree of neurological damage is bound to occur among the survivors of all situations which carry a high risk of perinatal mortality and that a high or even a known proportion of mental retardation can be ascribed to the non-lethal grades of reproductive difficulty. A large study reported by Buck (1968) indicates that the most common reproductive difficulties when occurring singly have no significant effect on children's intellectual status after age 5, with the one exception of pre-eclamptic toxemia of pregnancy, which caused some cognitive impairment. Most of the complications of pregnancy, it seems, must occur multiply to impair intellectual ability. It is as if the nervous system is sufficiently homeostatic to withstand certain unfavorable conditions if they occur singly.

Prematurity. The literature on the relationship of premature birth to the child's IQ is confusing and conflicting. Guilford (1967), in his recent book on *The Nature of Intelligence,* for example, concluded, as did Stoddard (1943), that prematurity has no effect on intelligence. Stott (1966), on the other hand, presents impressive evidence of very significant IQ decrements associated with prematurity. Probably the most thorough review of the subject I have found, by Kushlick (1966), helps to resolve these conflicting opinions. There is little question that prematurity has the strongest known relation to brain dysfunction of any reproductive factor, and many of the complications of pregnancy are strongly associated with the production of premature children. The crucial factor in prematurity, however, is not prematurity per se, but low birth-weight. Birth-weight apparently acts as a threshold variable with respect to intellectual impairment. All studies of birth-weight agree in showing that the incidence of babies weighing less than 5-1/2 lbs. increases from higher to lower social classes. But only about 1 percent of the total variance of birth-weight is accounted for by socioeconomic variables. Race (Negro versus white) has an effect on birth-weight independently of socioeconomic variables. Negro babies mature at a lower birth-weight than white babies (Naylor & Myrianthopoulos, 1967). If prematurity is defined as a condition in which birth-weight is under 5-1/2 lbs., the observed relationship between prematurity and depression of the IQ is due to the common factor of low social class. Kushlick (1966, p. 143) concludes that it is only among children having birth-weights under 3 lbs. that the mean IQ is lowered, independently of social class, and more in boys than in girls. The incidence of extreme subnormality is higher for children with birth-weights under 3 or 4 lbs. But when one does not count these extreme cases (IQs below 50), the effects of prematurity or low birth-weight—even as low as 3 lbs.—have a very weak relationship to children's IQs by the time they are of school age. The association between very low birth-weight and extreme mental subnormality raises the question of whether the low birth-weight causes the abnormality or whether the abnormality arises independently and causes the low birth-weight.

Prematurity and low birth-weight have a markedly higher incidence among Negroes than among whites. That birth-weight differences per se are not a predominant factor in Negro-white IQ differences, however, is suggested by the findings of a study which compared Negro and white premature children matched for birth-weight. The Negro children in all weight groups performed significantly less well on mental tests at 3 and 5 years of age than the white children of comparable birth-weight (Hardy, 1965, p. 51).

Genetic Predisposition to Prenatal Impairment. Dennis Stott (1960, 1966), a British psychologist, has adduced considerable evidence for the theory that impairments of the central nervous system occurring prenatally as a result of various stresses in pregnancy may not be the *direct* result of adverse intrauterine factors but may result *indirectly* from genetically determined mechanisms which are triggered by prenatal stress of one form or another.

Why should there exist a genetic mechanism predisposing to congenital impairments? Would not such genes, if they had ever existed, have been eliminated long ago through natural selection? It can be argued from considerable evidence in lower species of mammals observable by zoologists today that such a genetic mechanism may have had survival value for primitive man, but that the conditions of our present industrial society and advances in medical care have diminished the biological advantage of this mechanism for survival of the human species. The argument is that, because of the need to control population, there is a genetic provision within all species for multiple impairments, which are normally only potentialities, that can be triggered off by prenatal stress associated with high population density, such as malnutrition, fatigue from overexertion, emotional distress, infections, and the like. The resulting congenital impairment would tend to cut down the infant population, thereby relieving the pressure of population without appreciably reducing the functioning and efficiency of the young adults in the population. Stott (1966) has presented direct evidence of an association between stresses in the mother during pregnancy and later behavioral abnormalities and learning problems of the child in school. The imperfect correlation between such prenatal stress factors and signs of congenital impairment suggests that there are individual differences in genetic predisposition to prenatal impairment. The hypothesis warrants further investigation. The prenatal environment could be a much more important source of later IQ variance for some children than for others.

Mother-Child Rh Incompatibility. The *Rh* blood factor can involve possible brain damaging effects in a small proportion of pregnancies where the fetus is *Rh*-positive and the mother is *Rh*-negative. (*Rh*-negative has a frequency of 15 percent in the white and 7 percent in the Negro population.) The mother-child *Rh* incompatibility produces significant physical ill effects in only a fraction of cases and increases in importance in pregnancies beyond the first. The general finding of slightly lower IQs in second and later born children could be related to *Rh* incompatibility or to similar, but as yet undiscovered, mother-child biological incompatibilities. This is clearly an area greatly in need of pioneering research.

Nutrition. Since the human brain attains 70 percent of its maximum adult weight in the first year after birth, it should not be surprising that prenatal and infant nutrition can have significant effects on brain development. Brain growth is largely a process of protein synthesis. During the prenatal period and the first postnatal year the brain normally absorbs large amounts of protein nutrients and grows at the average rate of 1 to 2 milligrams per minute (Stoch & Smythe, 1963; Cravioto, 1966).

Severe undernutrition before two or three years of age, especially a lack of proteins and the vitamins and minerals essential for their anabolism, results in lowered intelligence. Stoch and Smythe (1963) found, for example, that extremely malnourished South African colored children were some 20 points lower in IQ than children of similar parents who had not suffered from malnutrition. The difference between the undernourished group and the control group in DQ and IQ over the age range from 1 year to 8 years was practically constant. If undernutrition takes a toll, it takes it early, as shown by the lower DQs at 1 year and the absence of any increase in the decrement at later ages. Undernutrition occurring for the first time in older children seems to have no permanent effect. Severely malnourished war prisoners, for example, function intellectually at their expected level when they are returned to normal living conditions. The study by Stoch and Smythe, like several others (Cravioto, 1966; Scrimshaw, 1968), also revealed that the undernourished children had smaller stature and head circumference than the control children. Although there is no correlation between intelligence and head circumference in normally nourished children, there is a positive correlation between these factors in groups whose numbers suffer varying degrees of undernutrition early in life. Undernutrition also increases the correlation between intelligence and physical stature. These correlations provide us with an index which could aid the study of IQ deficits due to undernutrition in selected populations.

One of the most interesting and pronounced psychological effects of undernutrition is retardation in the development of cross-modal transfer or intersensory integration, which was earlier described as characterizing the essence of *g* (Scrimshaw, 1968).

The earlier the age at which nutritional therapy is instituted, of course, the more beneficial are its effects. But even as late as 2 years of age, a gain of as much as 18 IQ points was produced by nutritional improvements in a group of extremely undernourished children. After 4 years of age, however, nutritional therapy effected no significant change in IQ (Cravioto, 1966, p. 82).

These studies were done in countries where extreme undernutrition is not uncommon. Such gross nutritional deprivation is rare in the United States. But there is at least one study which shows that some undetermined proportion of the urban population in the United States might benefit substantially with respect to intellectual development by improved nutrition. In New York City, women of low socioeconomic status were given vitamin and mineral supplements during pregnancy. These women gave birth to children who, at four years of age, averaged 8 points higher in IQ than a control group of children whose mothers had been given placebos during pregnancy (Harrell, Woodyard, & Gates, 1955). Vitamin and mineral supplements are, of course, beneficial in this way only when they remedy an existing deficiency.

Birth Order. Order of birth contributes a significant proportion of the variance in mental ability. On the average, first-born children are superior in almost every way, mentally and physically. This is the consistent finding of many studies (Altus, 1966) but as yet the phenomenon remains unexplained. (Rimland [1964, pp. 140-143] has put forth some interesting hypotheses to explain the superiority of the first-born.) Since the first-born effect is found throughout all social classes in many countries and has shown up in studies over the past 80 years (it was first noted by Galton), it is probably a biological rather than a social-psychological phenomenon. It is almost certainly not a genetic effect. (It would tend to make for slightly lower estimates of heritability based on sibling comparisons.) It is one of the sources of environmental variance in ability without any significant postnatal environmental correlates. No way is known for giving later-born children the same advantage. The disadvantage of being later-born, however, is very slight and shows up conspicuously only in the extreme upper tail of the distribution of achievements. For example, there is a disproportionate number of first-born individuals whose biographies appear in *Who's Who* and in the *Encyclopedia Britannica*.

Social Class Differences in Intelligence

Social class (or socioeconomic status [SES]) should be considered as a factor separate from race. I have tried to avoid using the terms *social class* and *race* synonymously or interchangeably in my writings, and I observe this distinction here. Social classes completely cut across all racial groups. But different racial groups are disproportionately represented in different SES categories. Social class differences refer to a socioeconomic continuum *within* racial groups.

It is well known that children's IQs, by school age, are correlated with the socio-economic status of their parents. This is a world-wide phenomenon and has an extensive research literature going back 70 years. Half of all the correlations between SES and children's IQs reported in the literature fall between .25 and .50, with most falling in the region of .35 to .40. When school children are grouped by SES, the mean IQs of the groups vary over a range of one to two standard deviations (15 to 30 IQ points), depending on the method of status classification (Eells, et al., 1951). This relationship between SES and IQ constitutes one of the most substantial and least disputed facts in psychology and education.

The fact that intelligence is correlated with occupational status can hardly be surprising in any society that supports universal public education. The educational system and occupational hierarchy act as an intellectual "screening" process, far from perfect, to be sure, but discriminating enough to create correlations of the magnitude just reported. If each generation is roughly sorted out by these "screening" processes along an intelligence continuum, and if, as has already been pointed out, the phenotype-genotype correlation for IQ is of the order of .80 to .90, it is almost inevitable that this sorting process will make for genotypic as well as phenotypic differences among social classes. It is therefore most unlikely that groups differing in SES would not also differ, on the average, in their genetic endowment of intelligence. In reviewing the relevant evidence, the British geneticist, C. O. Carter (1966, p. 192) remarked, "Sociologists who doubt this show more ingenuity than judgment." Sociologist Bruce Eckland (1967) has elaborately spelled out the importance of genetic factors for understanding social class differences.

Few if any students of this field today would regard socioeconomic status per se as an environmental variable that primarily *causes* IQ differences. Intellectual differences between SES groups have hereditary, environmental, and interaction components. Environmental factors associated with SES differences apparently are not a major *independent* source of variance in intelligence. Identical twins separated in the first months of life and reared in widely differing social classes, for example, still show greater similarity in IQ than unrelated children reared together or than even siblings reared together (Burt, 1966). The IQs of children adopted in infancy show a much lower correlation with the SES of the adopting parents than do the IQs of children reared by their own parents (Leahy, 1935). The IQs of children who were reared in an orphanage from infancy and who had never known their biological parents show approximately the same correlation with their biological father's occupational status as found for children reared by their biological parents (.23 *vs* .24) (Lawrence, 1931). The correlation between the IQs of children adopted in infancy and the educational level of their biological mothers is close to that of children reared by their own mothers (.44), while the correlation between children's IQs and their adopting parents' educational level is close to zero (Honzik, 1957). Children of low and high SES show, on the average, an amount of regression from the parental IQ toward the mean of the general population that conforms to expectations from a simple polygenic model of the inheritance of intelligence (Burt, 1961). When siblings reared within the same family differ significantly in intelligence, those who are above the family average tend to move up the SES scale, and those who are below the family average tend to move down (Young & Gibson, 1965). It should also be noted that despite intensive efforts by psychologists, educators, and sociologists to devise tests intended to eliminate SES differences in measured intelligence, none of these efforts has succeeded (Jensen, 1968c). Theodosius Dobzhansky (1968a, p. 33), a geneticist, states that "There exist some occupations or functions for which only extreme genotypes are suitable." But surely this is not an all-or-nothing affair, and we would expect by the same reasoning that many different occupational skills, and not just those that are the most extreme, would favor some genotypes more than others. To be sure, genetic factors become more important at the extremes. Some minimal level of ability is required for learning most skills. But while you can teach almost anyone to play chess, or the piano, or to conduct an orchestra, or to write prose, you cannot teach everyone to be a Capablanca, a Paderewski, a Toscanini, or a Bernard Shaw. In a society that values and rewards individual talent and merit, genetic factors inevitably take on considerable importance.

SES differences, and race differences as well, are manifested not only as differences between group means, but also as differences in variance and in patterns of correlations among various mental abilities, even on tests which show no *mean* differences between SES groups (Jensen, 1968b).

Another line of evidence that SES IQ differences are not a superficial phenomenon is the fact of a negative correlation between SES and Developmental Quotient (DQ) (under two years of age) and an increasing positive correlation between SES and IQ (beyond two years of age), as shown in Figure 14 from a study by Nancy Bayley (1966). (All subjects in this study are Caucasian.) This relationship is especially interesting in view of the finding of a number of studies that there is a negative correlation between DQ and later IQ, an effect which is much more pronounced in boys than in girls and involves the motor more than the attentional-cognitive aspects of the DQ (Bayley, 1965b). Figure 14 shows that on

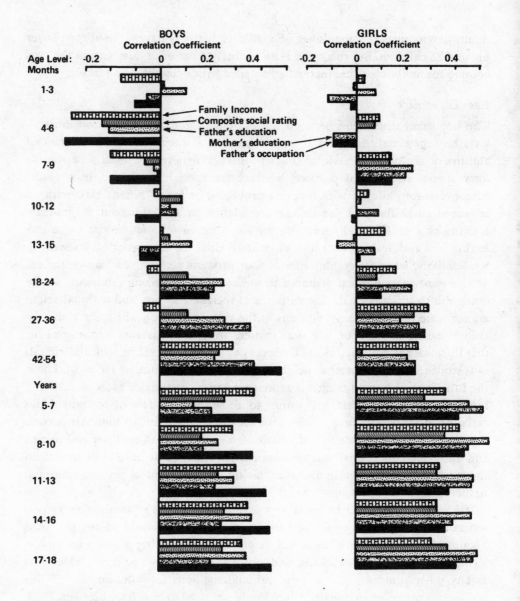

FIGURE 14.

Correlations between children's mental test scores, at 1 month to 18 years, and five indicators of parents' socioeconomic status at the time the children were born. (Bayley, 1966.)

infant developmental scales, lower SES children actually have a "head start" over higher SES children. But this trend is increasingly reversed at later ages as the tests become less motoric and are increasingly loaded with a cognitive or *g* factor.

Race Differences

The important distinction between the *individual* and the *population* must always be kept clearly in mind in any discussion of racial differences in mental abilities or any other behavioral characteristics. Whenever we select a person for some special educational purpose, whether for special instruction in a grade-school class for children with learning problems, or for a "gifted" class with an advanced curriculum, or for college attendance, or for admission to graduate training or a professional school, we are selecting an *individual*, and we are selecting him and dealing with him as an individual for reasons of his individuality. Similarly, when we employ someone, or promote someone in his occupation, or give some special award or honor to someone for his accomplishments, we are doing this to an individual. The variables of social class, race, and national origin are correlated so imperfectly with any of the valid criteria on which the above decisions should depend, or, for that matter, with any behavioral characteristic, that these background factors are irrelevant as a basis for dealing with individuals —as students, as employees, as neighbors. Furthermore, since, as far as we know, the full range of human talents is represented in all the major races of man and in all socioeconomic levels, it is unjust to allow the mere fact of an individual's racial or social background to affect the treatment accorded to him. All persons rightfully must be regarded on the basis of their individual qualities and merits, and all social, educational, and economic institutions must have built into them the mechanisms for insuring and maximizing the treatment of persons according to their individual behavior.

If a society completely believed and practiced the ideal of treating every person as an individual, it would be hard to see why there should be any problems about "race" per se. There might still be problems concerning poverty, unemployment, crime, and other social ills, and, given the will, they could be tackled just as any other problems that require rational methods for solution. But if this philosophy prevailed in practice, there would not need to be a "race problem."

The question of *race* differences in intelligence comes up not when we deal with individuals as individuals, but when certain identifiable *groups* or subcultures within the society are brought into comparison with one another *as groups or populations*. It is only when the groups are disproportionately represented in what are commonly perceived as the most desirable and the least desirable social and occupational roles in a society that the question arises concerning average differences among groups. Since much of the current thinking behind civil rights, fair employment, and equality of educational opportunity appeals to the fact that there is a disproportionate representation of different racial groups in the various levels of the educational, occupational, and socioeconomic hierarchy, we are forced to examine all the possible reasons for this inequality among racial groups in the attainments and rewards generally valued by all groups within our society. To what extent can such inequalities be attributed to unfairness in society's multiple selection processes? ("Unfair" meaning that selection is influenced by intrinsically irrelevant criteria, such as skin color, racial or national origin, etc.) And to what extent are these inequalities attributable to really relevant selection criteria which apply equally to all individuals but at the same time select disproportionately between some racial groups because there exist, in fact, real average differences among the groups—differences in the population distributions of those characteristics which are indisputably relevant to educational and occupational performance? This is certainly one of the most important questions confronting our nation today. The answer, which can be found only through unfettered research, has enormous consequences for the welfare of all, particularly of minorities whose plight is now in the foreground of public attention. A preordained, doctrinaire stance with regard to this issue hinders the achievement of a scientific understanding of the problem. To rule out of court, so to speak, any reasonable hypotheses on purely ideological grounds is to argue that static ignorance is preferable to increasing our knowledge of reality. I strongly disagree with those who believe in searching for the truth by scientific means only under certain circumstances and eschew this course in favor of ignorance under other circumstances, or who believe that the results of inquiry on some subjects cannot be entrusted to the public but should be kept the guarded possession of a scientific elite. Such attitudes, in my opinion, represent a danger to free inquiry and, consequently, in the long run, work to the disadvantage of society's general welfare. "No holds barred" is the best formula for scientific inquiry. One does not decree beforehand which phenomena cannot be studied or which questions cannot be answered.

Genetic Aspects of Racial Differences. No one, to my knowledge, questions the role of environmental factors, including influences from past history, in determining at least some of the variance between racial groups in standard measures of intelligence, school performance, and occupational status. The current literature on the culturally disadvantaged abounds with discussion—some of it factual, some of it fanciful—of how a host of environmental factors depresses cognitive development and performance. I recently co-edited a book which is largely concerned with the environmental aspects of disadvantaged minorities (Deutsch, Katz, & Jensen, 1968). But the possible importance of genetic factors in racial behavioral differences has been greatly ignored, almost to the point of being a tabooed subject, just as were the topics of venereal disease and birth control a generation or so ago.

My discussions with a number of geneticists concerning the question of a genetic basis of differences among races in mental abilities have revealed to me a number of rather consistently agreed-upon points which can be summarized in general terms as follows: Any groups which have been geographically or socially isolated from one another for many generations are practically certain to differ in their gene pools, and consequently are likely to show differences in any phenotypic characteristics having high heritability. This is practically axiomatic, according to the geneticists with whom I have spoken. Races are said to be "breeding populations," which is to say that matings within the group have a much higher probability than matings outside the group. Races are more technically viewed by geneticists as populations having different distributions of gene frequencies. These genetic differences are manifested in virtually every anatomical, physiological, and biochemical comparison one can make between representative samples of identifiable racial groups (Kuttner, 1967). There is no reason to suppose that the brain should be exempt from this generalization. (Racial differences in the relative frequencies of various blood constituents have probably been the most thoroughly studied so far.)

But what about behavior? If it can be measured and shown to have a genetic component, it would be regarded, from a genetic standpoint, as no different from other human characteristics. There seems to be little question that racial differences in genetically conditioned behavioral characteristics, such as mental abilities, should exist, just as physical differences. The real questions, geneticists tell me, are not whether there are or are not genetic racial differences that affect behavior, because there undoubtedly are. The proper questions to ask, from a scientific standpoint, are: What is the direction of the difference? What is the magnitude of the difference? And what is the significance of the difference—medically, socially, educationally, or from whatever standpoint that may be relevant to the characteristic in question? A difference is important only within a specific context. For example, one's blood type in the ABO system is unimportant until one needs a transfusion. And some genetic differences are apparently of no importance with respect to any context as far as anyone has been able to discover—for example, differences in the size and shape of ear lobes. The idea that all genetic differences have arisen or persisted only as a result of natural selection, by conferring some survival or adaptive benefit on their possessors, is no longer generally held. There appear to be many genetic differences, or polymorphisms, which confer no discernible advantages to survival.[6]

Negro Intelligence and Scholastic Performance. Negroes in the United States are disproportionately represented among groups identified as culturally or educationally disadvantaged. This, plus the fact that Negroes constitute by far the largest racial minority in the United States, has for many years focused attention on Negro intelligence. It is a subject with a now vast literature which has been quite recently reviewed by Dreger and Miller (1960, 1968) and by Shuey (1966), whose 578 page review is the most comprehensive, covering 382 studies. The basic data are well known: on the average, Negroes test about 1 standard deviation (15 IQ points) below the average of the white population in IQ, and this finding is fairly uniform across the 81 different tests of intellectual ability used in the studies reviewed by Shuey. This magnitude of difference gives a median overlap of 15 percent, meaning that 15 percent of the Negro population exceeds the white average. In terms of proportions of variance, if the numbers of Negroes and whites were equal, the differences *between* racial groups would account for 23 percent of the total variance, but—an important point—the differences *within* groups would account for 77 percent of the total variance. When gross socioeconomic level is controlled, the average difference reduces to about 11 IQ points (Shuey, 1966, p. 519), which, it should be recalled, is about the same spread as the average difference between siblings in the same family. So-called "culture-free" or "culture-fair" tests tend to give Negroes slightly lower scores, on the average, than more conventional IQ tests such as the Stanford-Binet and Wechsler scales. Also, as a group, Negroes perform somewhat more poorly on those subtests which tap abstract abilities. The majority of studies show that Negroes perform relatively better on verbal than on non-verbal intelligence tests.

In tests of scholastic achievement, also, judging from the massive data of the Coleman study (Coleman, et al., 1966), Negroes score about 1 standard devia-

---

[6] The most comprehensive and sophisticated discussion of the genic-behavior analysis of race differences that I have found is by Spuhler and Lindzey (1967).

tion (SD) below the average for whites and Orientals and considerably less than 1 SD below other disadvantaged minorities tested in the Coleman study—Puerto Rican, Mexican-American, and American Indian. The 1 SD decrement in Negro performance is fairly constant throughout the period from grades 1 through 12.

Another aspect of the distribution of IQs in the Negro population is their lesser variance in comparison to the white distribution. This shows up in most of the studies reviewed by Shuey. The best single estimate is probably the estimate based on a large normative study of Stanford-Binet IQs of Negro school children in five Southeastern states, by Kennedy, Van De Riet, and White (1963). They found the SD of Negro children's IQs to be 12.4, as compared with 16.4 in the white normative sample. The Negro distribution thus has only about 60 percent as much variance (i.e., SD^2) as the white distribution.

There is an increasing realization among students of the psychology of the disadvantaged that the discrepancy in their average performance cannot be completely or directly attributed to discrimination or inequalities in education. It seems not unreasonable, in view of the fact that intelligence variation has a large genetic component, to hypothesize that genetic factors may play a part in this picture. But such an hypothesis is anathema to many social scientists. The idea that the lower average intelligence and scholastic performance of Negroes could involve, not only environmental, but also genetic, factors has indeed been strongly denounced (e.g., Pettigrew, 1964). But it has been neither contradicted nor discredited by evidence.

The fact that a reasonable hypothesis has not been rigorously proved does not mean that it should be summarily dismissed. It only means that we need more appropriate research for putting it to the test. I believe such definitive research is entirely possible but has not yet been done. So all we are left with are various lines of evidence, no one of which is definitive alone, but which, viewed all together, make it a not unreasonable hypothesis that genetic factors are strongly implicated in the average Negro-white intelligence difference. The preponderance of the evidence is, in my opinion, less consistent with a strictly environmental hypothesis than with a genetic hypothesis, which, of course, does not exclude the influence of environment or its interaction with genetic factors.

We can be accused of superficiality in our thinking about this issue, I believe, if we simply dismiss a genetic hypothesis without having seriously thought about the relevance of typical findings such as the following:

Failure to Equate Negroes and Whites in IQ and Scholastic Ability. No one has yet produced any evidence based on a properly controlled study to show that representative samples of Negro and white children can be equalized in intellectual ability through statistical control of environment and education.

Socioeconomic Level and Incidence of Mental Retardation. Since in no category of socioeconomic status (SES) are a majority of children found to be retarded in the technical sense of having an IQ below 75, it would be hard to claim that the degree of environmental deprivation typically associated with lower-class status could be responsible for this degree of mental retardation. An IQ less than 75 reflects more than a lack of cultural amenities. Heber (1968) has estimated on the basis of existing evidence that IQs below 75 have a much higher incidence among Negro than among white children at every level of socioeconomic status, as shown in Table 3. In the two highest SES categories the estimated proportions of Negro and white children with IQs below 75, are in the ratio of 13.6 to 1. If

TABLE 3

*Estimated Prevalence of Children With IQs Below 75, by
Socioeconomic Status (SES) and Race Given as Percentages
(Heber, 1968)*

SES	White	Negro
High 1	0.5	3.1
2	0.8	14.5
3	2.1	22.8
4	3.1	37.8
Low 5	7.8	42.9

environmental factors were mainly responsible for producing such differences, one should expect a lesser Negro-white discrepancy at the upper SES levels. Other lines of evidence also show this not to be the case. A genetic hypothesis, on the other hand, would predict this effect, since the higher SES Negro offspring would be regressing to a lower population mean than their white counterparts in SES, and consequently a larger proportion of the lower tail of the distribution of genotypes for Negroes would fall below the value that generally results in phenotypic IQs below 75.

A finding reported by Wilson (1967) is also in line with this prediction. He obtained the mean IQs of a large representative sample of Negro and white children in a California school district and compared the two groups within each of four social class categories: (1) professional and managerial, (2) white collar, (3) skilled and semiskilled manual, and (4) lower class (unskilled, unemployed, or welfare recipients). The mean IQ of Negro children in the first category was 15.5 points below that of the corresponding white children in SES category 1. But the Negro mean for SES 1 was also 3.9 points below the mean of white children in SES category 4. (The IQs of white children in SES 4 presumably have "regressed" upward toward the mean of the white population.)

Wilson's data are not atypical, for they agree with Shuey's (1966, p. 520) summarization of the total literature up to 1965 on this point. She reports that in all the studies which grouped subjects by SES, upper-status Negro children average 2.6 IQ points *below* the low-status whites. Shuey comments: "It seems improbable that upper and middle-class colored children would have no more culture opportunities provided them than white children of the lower and lowest class."

Duncan (1968, p. 69) also has presented striking evidence for a much greater "regression-to-the-mean" (from parents to their children) for high status occupations in the case of Negroes than in the case of whites. None of these findings is at all surprising from the standpoint of a genetic hypothesis, of which an intrinsic feature is Galton's "law of filial regression." While the data are not necessarily inconsistent with a possible environmental interpretation, they do seem more puzzling in terms of strictly environmental causation. Such explanations often seem intemperately strained.

Inadequacies of Purely Environmental Explanations. Strictly environmental explanations of group differences tend to have an ad hoc quality. They are usually plausible for the situation they are devised to explain, but often they have little generality across situations, and new ad hoc hypotheses have to be continually devised. Pointing to environmental differences between groups is never sufficient in itself to infer a causal relationship to group differences in intelligence. To take just one example of this tendency of social scientists to attribute lower intelligence and scholastic ability to almost any environmental difference that seems handy, we can look at the evidence regarding the effects of "father absence." Since the father is absent in a significantly larger proportion of Negro than of white families, the factor of "father absence" has been frequently pointed to in the literature on the disadvantaged as one of the causes of Negroes' lower performance on IQ tests and in scholastic achievement. Yet the two largest studies directed at obtaining evidence on this very point—the only studies I have seen that are methodologically adequate—both conclude that the factor of "father absence" versus "father presence" makes no independent contribution to variance in intelligence or scholastic achievement. The sample sizes were so large in both of these studies that even a very slight degree of correlation between father-absence and the measures of cognitive performance would have shown up as statistically significant. Coleman (1966, p. 506) concluded: "Absence of a father in the home did not have the anticipated effect on ability scores. Overall, pupils without fathers performed at approximately the same level as those with fathers—although there was some variation between groups" (groups referring to geographical regions of the U.S.). And Wilson (1957, p. 177) concluded from his survey of a California school district: "Neither our own data nor the preponderance of evidence from other research studies indicate that father presence or absence, *per se*, is related to school achievement. While broken homes reflect the existence of social and personal problems, and have some consequence for the development of personality, broken homes do not have any systematic effect on the overall level of school success."

The nationwide Coleman study (1966) included assessments of a dozen environmental variables and socioeconomic indices which are generally thought to be major sources of environmental influence in determining individual and group differences in scholastic performance—such factors as: reading material in the home, cultural amenities in the home, structural integrity of the home, foreign language in the home, preschool attendance, parents' education, parents' educational desires for child, parents' interest in child's school work, time spent on homework, child's self-concept (self-esteem), and so on. These factors are all correlated—in the expected direction—with scholastic performance within each of the racial or ethnic groups studied by Coleman. Yet, interestingly enough, they are not systematically correlated with differences *between* groups. For example, by far the most environmentally disadvantaged groups in the Coleman study are the American Indians. On every environmental index they average *lower* than the Negro samples, and overall their environmental rating is about as far below the Negro average as the Negro rating is below the white average. (As pointed out by Kuttner [1968, p. 707], American Indians are much more disadvantaged than Negroes, or any other minority groups in the United States, on a host of other factors not assessed by Coleman, such as income, unemployment, standards of health care, life expectancy, and infant mortality.) Yet the American Indian ability and achievement test scores average about half a standard deviation higher than the scores of Negroes. The differences were in favor of the Indian children on each of the four tests used by Coleman: non-verbal intelligence, verbal intelligence, reading comprehension, and math achievement. If the environmental factors assessed by Coleman are the major determinants of Negro-white differences that many social scientists have claimed they are, it is hard to see why such factors should act in reverse fashion in determining differences between Negroes and Indians, especially in view of the fact that *within* each group the factors are significantly correlated in the expected direction with achievement.

Early Developmental Differences. A number of students of child development have noted the developmental precocity of Negro infants, particularly in motoric behavior. Geber (1958) and Geber and Dean (1957) have reported this precocity also in African infants. It hardly appears to be environmental, since it is evident in nine-hour-old infants. Cravioto (1966, p. 78) has noted that the Gesell tests of infant behavioral development, which are usually considered suitable only for children over four weeks of age, "can be used with younger African, Mexican, and Guatemalan infants, since their development at two or three weeks is similar to that of Western European infants two or three times as old." Bayley's (1965a) study of a representative sample of 600 American Negro infants up to 15 months of age, using the Bayley Infant Scales of Mental and Motor Development, also found Negro infants to have significantly higher scores than white infants in their first year. The difference is largely attributable to the motor items in the Bayley test. For example, about 30 percent of white infants as compared with about 60 percent of Negro infants between 9 and 12 months were able to "pass" such tests as "pat-a-cake" muscular coordination, and ability to walk with help, to stand alone, and to walk alone. The highest scores for any group on the Bayley scales that I have found in my search of the literature were obtained by Negro infants in the poorest sections of Durham, North Carolina. The older siblings of these infants have an average IQ of about 80. The infants up to 6 months of age, however, have a Developmental Motor Quotient (DMQ) nearly one standard deviation above white norms and a Developmental IQ (i.e., the non-motor items of the Bayley scale) of about half a standard deviation above white norms (Durham Education Improvement Program, 1966-67, a, b).

The DMQ, as pointed out previously, correlates negatively in the white population with socioeconomic status and with later IQ. Since lower SES Negro and white school children are more alike in IQ than are upper SES children of the two groups (Wilson, 1967), one might expect greater DMQ differences in favor of Negro infants in high socioeconomic Negro and white samples than in low socioeconomic samples. This is just what Walters (1967) found. High SES Negro infants significantly exceeded whites in total score on the Gesell developmental schedules at 12 weeks of age, while low SES Negro and white infants did not differ significantly overall. (The only difference, on a single subscale, favored the white infants.)

It should also be noted that developmental quotients are usually depressed by adverse prenatal, perinatal, and postnatal complications such as lack of oxygen, prematurity, and nutritional deficiency.

Another relationship of interest is the finding that the negative correlation between DMQ and later IQ is higher in boys than in girls (Bayley, 1966, p. 127). Bronfenbrenner (1967, p. 912) cites evidence which shows that Negro boys perform relatively less well in school than Negro girls; the sex difference is much greater than is found in the white population. Bronfenbrenner (1967, p. 913) says, "It is noteworthy that these sex differences in achievement are observed among Southern as well as Northern Negroes, are present at every socioeconomic level, and tend to increase with age."

Physiological Indices. The behavioral precocity of Negro infants is also paralleled by certain physiological indices of development. For example, x-rays show that bone development, as indicated by the rate of ossification of cartilege, is more advanced in Negro as compared with white babies of about the same socioeconomic background, and Negro babies mature at a lower birth-weight than white babies (Naylor & Myrianthopoulos, 1967, p. 81).

It has also been noted that brain wave patterns in African newborn infants show greater maturity than is usually found in the European newborn child (Nilson & Dean, 1959). This finding especially merits further study, since there is evidence that brain waves have some relationship to IQ (Medical World News, 1968), and since at least one aspect of brain waves—the visually evoked potential —has a very significant genetic component, showing a heritability of about .80 (uncorrected for attenuation) (Dustman & Beck, 1965).

Magnitude of Adult Negro-White Differences. The largest sampling of Negro and white intelligence test scores resulted from the administration of the Armed Forces Qualification Test (AFQT) to a national sample of over 10 million men between the ages of 18 and 26. As of 1966, the overall failure rate for Negroes was 68 percent as compared with 19 percent for whites (*U.S. News and World Report*, 1966). (The failure cut-off score that yields these percentages is roughly equivalent to a Stanford-Binet IQ of 86.) Moynihan (1965) has estimated that during the same period in which the AFQT was administered to these large representative samples of Negro and white male youths, approximately one-half of Negro families could be considered as middle-class or above by the usual socioeconomic criteria. So even if we assumed that all of the lower 50 percent of Negroes on the SES scale failed the AFQT, it would still mean that at least 36 percent of the middle SES Negroes failed the test, a failure rate almost twice as high as that of the white population for all levels of SES.

Do such findings raise any question as to the plausibility of theories that postulate exclusively environmental factors as sufficient causes for the observed differences?

Why Raise Intelligence?

If the intelligence of the whole population increased and our IQ tests were standardized anew, the mean IQ would again be made equal to 100, which, by definition, is the average for the population. Thus, in order to speak sensibly of raising intelligence we need an absolute frame of reference, and for simplicity's sake we will use the *present* distribution of IQ as our reference scale. Then it will not be meaningless to speak of the average IQ of the population shifting to values other than 100.

Would there be any real advantage to shifting the entire distribution of intelligence upward? One way to answer this question is to compare the educational attainments of children in different schools whose IQ distributions center around means of, say, 85, 100, and 115. As pointed out earlier, there is a relationship between educational attainments and the occupations that are open to individuals on leaving school. Perusal of the want-ads in any metropolitan newspaper reveals that there are extremely few jobs advertised which are suitable to the level of education and skills typically found below IQs of 85 or 90, while we see day after day in the want-ads hundreds of jobs which call for a level of education and skills typically found among school graduates with IQs above 110. These jobs go begging to be filled. The fact is, there are not nearly enough minimally qualified persons to fill them.

One may sensibly ask the question whether our collective national intelligence is adequate to meet the growing needs of our increasingly complex industrial society. In a bygone era, when the entire population's work consisted almost completely of gathering or producing food by primitive means, there was little need for a large number of persons with IQs much above 100. Few of the jobs that had to be done at that time required the kinds of abstract intelligence and academic training which are now in such seemingly short supply in relation to the demand in our modern society. For many years the criterion for mental retardation was an IQ below 70. In recent years the National Association for Mental Retardation has raised the criterion to an IQ of 85, since an increasing proportion of persons of more than 1 standard deviation below the average in IQ are unable to get along occupationally in today's world. Persons with IQs of 85 or less are finding it increasingly difficult to get jobs, any jobs, because they are unprepared, for whatever reason, to do the jobs that need doing in this industrialized, technological economy. Unless drastic changes occur—in the population, in educational outcomes, or in the whole system of occupational training and selection—it is hard to see how we can avoid an increase in the rate of the so-called "hard-core" unemployed. It takes more knowledge and cleverness to operate, maintain, or repair a tractor than to till a field by hand, and it takes more skill to write computer programs than to operate an adding machine. And apparently the trend will continue.

It has been argued by Harry and Margaret Harlow that "human beings in our world today have no more, or little more, than the absolute minimal intellectual endowment necessary for achieving the civilization we know today" (Harlow & Harlow, 1962, p. 34). They depict where we would probably be if man's average genetic endowment for intelligence had never risen above the level corresponding to IQ 75: "... the geniuses would barely exceed our normal or average level; comparatively few would be equivalent in ability to our average high school graduates. There would be no individuals with the normal intellectual capacities essential for making major discoveries, and there could be no civilization as we know it."

It may well be true that the kind of ability we now call intelligence was needed in a certain percentage of the human population for our civilization to have arisen. But while a small minority—perhaps only one or two percent—of highly gifted individuals were needed to advance civilization, the vast majority were able to assimilate the consequences of these advances. It may take a Leibnitz or a Newton to invent the calculus, but almost any college student can learn it and use it.

Since intelligence (meaning g) is not the whole of human abilities, there may be some fallacy and some danger in making it the *sine qua non* of fitness to play a productive role in modern society. We should not assume certain ability requirements for a job without establishing these requirements as a fact. How often do employment tests, Civil Service examinations, the requirement of a high school diploma, and the like, constitute hurdles that are irrelevant to actual performance on the job for which they are intended as a screening device? Before going overboard in deploring the fact that disadvantaged minority groups fail to clear many of the hurdles that are set up for certain jobs, we should determine whether the educational and mental test barriers that stand at the entrance to many of these employment opportunities are actually relevant. They may be relevant only in the correlational sense that the test predicts success on the job, in which case we should also know whether the test measures the ability actually required on the job or measures only characteristics that happen to be correlated with some third factor which is really essential for job performance. Changing people in terms of the really essential requirements of a given job may be much more feasible than trying to increase their abstract intelligence or level of performance in academic subjects so that they can pass irrelevant tests.

IQ Gains from Environmental Improvement

As was pointed out earlier, since the environment acts as a threshold variable with respect to IQ, an overall increase in IQ in a population in which a great majority are above the threshold, such that most of the IQ variance is due to heredity, could not be expected to be very large if it had to depend solely upon improving the environment of the economically disadvantaged. This is not to say that such improvement is not to be desired for its own sake or that it would not boost the educational potential of many disadvantaged children. An unrealistically high upper limit of what one could expect can be estimated from figures given by Schwebel (1968, p. 210). He estimates that 26 percent of the children in the population can be called environmentally deprived. He estimates the frequencies of their IQs in each portion of the IQ scale; their distribution is skewed, with higher frequencies in the lower IQ categories and an overall mean IQ of 90. Next, he assumes we could add 20 points to each deprived child's IQ by giving him an abundant environment. (The figure of 20 IQ points comes from Bloom's [1964, p. 89] estimate that the effect of extreme environments on intelligence is about 20 IQ points.) The net effect of this 20-point boost in the IQ of every deprived child would be an increase in the population's IQ from 100 to 105. But this seems to be an unrealistic fantasy. For if it were true that the IQs of the deprived group could be raised 20 points by a good environment, and if Schwebel's estimate of 26 percent correctly represents the incidence of deprivation, then the deprived children would be boosted to an average IQ of 110, which is 7 points higher than the mean of 103 for the non-deprived population! There is no reason to believe that the IQs of deprived children, given an environment of abundance, would rise to a higher level than the already privileged children's IQs. The overall boost in the population IQ would probably be more like 1 or 2 IQ points rather than 5. (Another anomaly of Schwebel's "analysis" is that after a 20-point IQ boost is granted to the deprived segment of the population, the only persons left in the mentally retarded range are the non-deprived, with 7 percent of them below IQ 80 as compared with zero percent of the deprived!)

Fewer persons, however, are seriously concerned about whether or not we could appreciably boost the IQ of the population as a whole. A more feasible and urgent goal is to foster the educational and occupational potential of the disadvantaged segment of the population. The pursuit of this aim, of course, must involve advances not only in education, but in public health, in social services, and in welfare and employment practices. In considering all feasible measures, one must also take inventory of forces that may be working against the accomplishment of amelioration. We should not overlook the fact that social and economic conditions not only have direct environmental effects, but indirectly can have biological consequences as well, consequences that could oppose attempts to improve the chances of the disadvantaged to assume productive roles in society.

Possible Dysgenic Trends

In one large midwestern city it was found that one-third of all the children in classes for the mentally retarded (IQ less than 75) came from one small area of the city comprising only five percent of the city's population (Heber, 1968). A representative sample of 88 mothers having at least one school-age child in this neighborhood showed an average of 7.6 children per mother. In families of 8 or more, nearly half the children over 12 years of age had IQs below 75 (Heber, Dener, & Conry, 1968). The authors note that not all low SES families contributed equally to the rate of mental retardation in this area; certain specifiable families had a greatly disproportionate number of retarded children. Mothers with IQs below 80, for example, accounted for over 80 percent of the children with IQs under 80. Completely aside from the hereditary implications, what does this mean in view of studies of foster children which show that the single most important factor in the child's *environment* with respect to his intellectual development is his foster mother's IQ? This variable has been shown to make the largest *independent* contribution to variance in children's IQs of any environmental factor (Burks, 1928). If the children in the neighborhoods studied by

91

Heber, which are typical of the situation in many of our large cities, have the great disadvantage of deprived environments, is it inappropriate to ask the same question that Florence Goodenough (1940, p. 329) posed regarding causal factors in retarded Tennessee mountain children: "*Why* are they so deprived?" When a substantial proportion of the children in a community suffer a deplorable environment, one of the questions we need to answer is who creates their environment? Does not the genetic $\times$ environment interaction work both ways, the genotype to some extent making its own environment and that of its progeny?

In reviewing evidence from foster home studies on environmental amelioration of IQs below 75 (the range often designated as indicating cultural-familial retardation) Heber, Dever, and Conry (1968, p. 17) state: "The conclusion that changes in the living environment can cause very large increments in IQ *for the cultural-familial retardate* is not warranted by these data."

What is probably the largest study every made of familial influences in mental retardation (defined in this study as IQ less than 70) involved investigation of more than 80,000 relatives of a group of mentally retarded persons by the Dight Institute of Genetics, University of Minnesota (Reed & Reed, 1965). From this large-scale study, Sheldon and Elizabeth Reed estimated that about 80 percent of mentally retarded (IQ less than 70) persons in the United States have a retarded parent or a normal parent who has a retarded sibling. The Reeds state: "One inescapable conclusion is that the transmission of mental retardation from parent to child is by far the most important *single* factor in the persistence of this social misfortune" (p. 48). "The transmission of mental retardation from one generation to the next, should, therefore, receive much more critical attention than it has in the past. It seems fair to state that this problem has been largely ignored on the assumption that if our social agencies function better, that if everyone's environment were improved sufficiently, then mental retardation would cease to be a major problem" (p. 77).

An interesting sidelight of the Reeds' study is the finding that in a number of families in which one or both parents had IQs below 70 and in which the environment they provided their children was deplorably deprived, there were a few children of average and superior IQ (as high as 130 or above) and superior scholastic performance. From a genetic standpoint the occurrence of such children would be expected. It is surprising from a strictly environmental standpoint. But, even though some proportion of the children of retarded parents are obviously intellectually well endowed, who would wish upon them the kind of environment typically provided by retarded parents? An investigation conducted in Denmark concluded that "...it is a very severe psychical trauma for a normally gifted child to grow up in a home where the mother is mentally deficient" (Jepsen & Bredmose, 1956, p. 209). Have we thought sufficiently of the rights of children—of their right to be born with fair odds against being mentally retarded, not to have a retarded parent, and with fair odds in favor of having the genetic endowment needed to compete on equal terms with the majority of persons in society? Can we reasonably and humanely oppose such rights of millions of children as yet not born?

Is Our National IQ Declining? It has long been known that there is a substantial negative correlation (averaging about -.30 in various studies) between intelligence and family size and between social class and family size (Anastasi, 1956). Children with many siblings, on the average, have lower IQs than children in small families, and the trend is especially marked for families of more than five (Gottesman, 1968). This fact once caused concern in the United States, and even more so in Britain, because of its apparent implication of a declining IQ in the population. If more children are born to persons in the lower half of the intelligence distribution, one would correctly predict a decline in the average IQ of the population. In a number of large-scale studies addressed to the issue in Britain and the United States some 20 years ago, no evidence was found for a general decline in IQ (Duncan, 1952). The paradox of the apparent failure of the genetic prediction to be manifested was resolved to the satisfaction of most geneticists by three now famous studies, one by Higgins, Reed, and Reed (1962), the others by Bajema (1963, 1966). All previous analyses had been based on IQ comparisons of children having different numbers of siblings, and this was their weakness. The data needed to answer the question properly consist of the average number of children born to *all* individuals at every level of IQ. It was found in the three studies that if persons with very low IQs married and had children, they typically had a large number of children. *But*—it was also found that relatively few persons in the lower tail of the IQ distribution ever married or produced children, and so their reproduction rate is more than counterbalanced by persons at the upper end of the IQ scale, nearly all of whom marry and have children. The data of these studies are shown in Figure 15.

In my opinion these studies are far from adequate to settle this issue and thus do not justify complacency. They cannot be generalized much beyond the particular generation which the data represent or to other than the white population on which these studies were based. The population sampled by Bajema

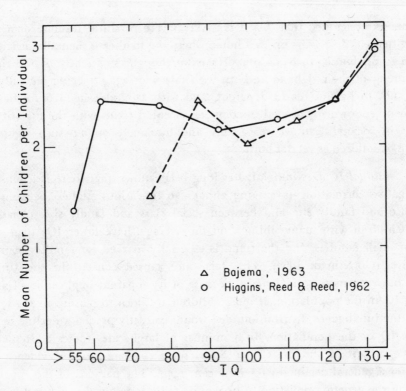

FIGURE 15.

Mean number of children per adult individual (including those who are child-less) at each level of IQ, in two samples of white American populations. Note in each sample the bimodal relationship between fertility and IQ.

(1963, 1966), for example, consisted of native-born American whites, predominantly Protestant, with above-average educational attainments, living all or most of their lives in an urban environment, and having most of their children before World War II. Results from a study of this population cannot be confidently generalized to other, quite dissimilar segments of our national population. The relationship between reproductive rate and IQ found by Bajema and by Higgins et al. may very well not prevail in every population group. Thus the evidence to date has not nullified the question of whether dysgenic trends are operating in some sectors.

If this conclusion is not unwarranted, then our lack of highly relevant information on this issue with respect to our Negro population is deplorable, and no one should be more concerned about it than the Negro community itself. Certain census statistics suggest that there might be forces at work which could create and widen the genetic aspect of the average difference in ability between the Negro and white populations in the United States, with the possible consequence that the improvement of educational facilities and increasing equality of opportunity will have a *decreasing* probability of producing equal achievement or continuing gains in the Negro population's ability to compete on equal terms. The relevant statistics have been presented by Moynihan (1966). The differential birthrate, as a function of socioeconomic status, is greater in the Negro than in the white population. The data showing this relationship for one representative age group from the U.S. Census of 1960 are presented in Figure 16.

Negro middle- and upper-class families have fewer children than their white counterparts, while Negro lower-class families have more. In 1960, Negro women of ages 35 to 44 married to unskilled laborers had 4.7 children as compared with 3.8 for non-Negro women in the same situation. Negro women married to professional or technical workers had only 1.9 children as compared with 2.4 for white women in the same circumstances. Negro women with annual incomes below $2000 averaged 5.3 children. The poverty rate for families with 5 or 6 children is 3½ times as high as that for families with one or two children (Hill & Jaffe, 1966). That these figures have some relationship to intellectual ability is seen in the fact that 3 out of 4 Negroes failing the Armed Forces Qualification Test come from families of four or more children.

Another factor to be considered is average generation time, defined as the number of years it takes for the parent generation to reproduce its own number. This period is significantly less in the Negro than in the white population. Also, as noted in the study of Bajema (1966), generation length is inversely related to educational attainment and occupational status; therefore a group with shorter generation length is more likely subject to a possible dysgenic effect.

Much more thought and research should be given to the educational and social implications of these trends for the future. Is there a danger that current welfare policies, unaided by eugenic foresight, could lead to the genetic enslavement of a substantial segment of our population? The possible consequences of our failure seriously to study these questions may well be viewed by future generations as our society's greatest injustice to Negro Americans.

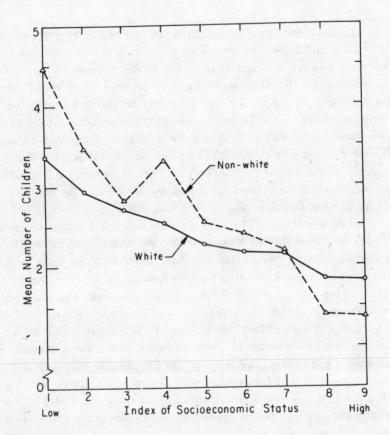

FIGURE 16.

Average number of children per woman 25 to 29 years of age, married once, with husband present, by race and socioeconomic status. From 1960 U.S. Census. (After Mitra, 1966.)

Intensive Educational Intervention

We began with mention of several of the major compensatory education programs and their general lack of success in boosting the scholastic performance of disadvantaged children. It has been claimed that such mammoth programs have not been adequately pinpointed to meeting specific, fine-grained cultural and cognitive needs of these children and therefore should not be expected to produce the gains that could result from more intensive and more carefully focused programs in which maximum cultural enrichment and instructional ingenuity are lavished on a small group of children by a team of experts.

The scanty evidence available seems to bear this out. While massive compensatory programs have produced no appreciable gains in intelligence or achievement (as noted on pp. 2-3), the majority of small-scale experiments in boosting the IQ and educational performance of disadvantaged children have produced significant gains. It is interesting that the magnitude of claimed gains generally decreases as one proceeds from reports in the popular press, to informal verbal reports heard on visits at research sites and in private correspondence, to papers read at meetings, to published papers without presentation of supporting data, and to published papers with supporting data. I will confine my review to some of the major studies in the last category.

First, some general observations.

Magnitude of Gains. The magnitude of IQ and scholastic achievement gains resulting from enrichment and cognitive stimulation programs authentically range between about 5 and 20 points for IQs, and between about one-half to two standard deviations for specific achievement measures (reading, arithmetic, spelling, etc.). Heber (1968) reviewed 29 intensive preschool programs for disadvantaged children and found they resulted in an average gain in IQ (at the time of children's leaving the preschool program) of between 5 and 10 points; the average gain was about the same for children whose initial IQs were below 90 as for those of 90 and above.

The amount of gain is related to several factors. The intensity and specificity of the instructional aspects of the program seem to make a difference. Ordinary nursery school attendance, with a rather diffuse enrichment program but with little effort directed at development of specific cognitive skills, generally results in a gain of 5 or 6 IQ points in typical disadvantaged preschoolers. If special cognitive training, especially in verbal skills, is added to the program, the average gain is about 10 points—slightly more or less depending on the amount of verbal content in the tests. Average gains rarely go above this, but when the program is extended beyond the classroom into the child's home, and there is intensive instruction in specific skills under short but highly attention-demanding daily sessions, as in the Bereiter-Engelmann program (1966), about a third of the children have shown gains of as much as 20 points.

Average gains of more than 10 or 15 points have not been obtained on any sizeable groups or been shown to persist or to be replicable in similar groups, although there have been claims that average gains of 20 or more points can be achieved by removing certain cultural and attitudinal barriers to learning. The actual evidence, however, warrants the caution expressed by Bereiter and Engelmann (1966, p. 7): " 'Miracle cures' of this kind are sometimes claimed to work with disadvantaged children, as when a child is found to gain 20 points or so in IQ after a few months of preschool experience. Such enormous gains, however, are highly suspect to anyone who is familiar with mental measurements. It is a fair guess that the child could have done as well on the first test except that he misinterpreted the situation, was frightened or agitated, or was not used to responding to instructions. Where genuine learning is concerned, enormous leaps simply do not occur, and leaps of any kind do not occur without sufficient cause."

The initial IQ on entering also has some effect, and this fact may be obscured if various studies are coarsely grouped. Bereiter and Engelmann (1966, p. 16), in analyzing results from eight different preschools for culturally disadvantaged children that followed traditional nursery school methods, concluded that the children's average gain in IQ is *half* the way from their initial IQ level to the normal level of 100. This rule was never more than 2 points in error for the studies reviewed. This same amount of IQ gain is generally noted in disadvantaged children during their first year in regular kindergarten (Brison, 1967, p. 8).

I have found no evidence of comparable gains in non-disadvantaged children. Probably the exceedingly meager gains in some apparently excellent preschool programs for the "disadvantaged" are attributable to the fact that the children in them did not come from a sufficiently deprived home background. Such can be the case when the children are admitted to the program on the basis of "self-selection" by their parents. Parents who seek out a nursery school or volunteer their children for an experimental preschool are more apt to have provided their children with a somewhat better environment than would be typical for a randomly selected group of disadvantaged children. This seems to have been the case in Martin Deutsch's intensive preschool enrichment program at the Institute of Developmental Studies in New York (Powledge, 1967). Both the experimental group (E) and the self-selected control groups (C_{ss}) were made up of Negro children from a poor neighborhood in New York City whose parents applied for their admission to the program. The E group received intensive educational attention in what is overall the most comprehensive and elaborate enrichment program I know of. The C_{ss} group, of course, received no enriched education.

The initial average Stanford-Binet IQs of the E and C_{ss} groups were 93.32 and 94.69, respectively. After two years in the enrichment program, the E group had a mean IQ of 95.53 and the C_{ss} group had 96.52. Both pre- and post-test differences are nonsignificant. The enrichment program continued for a third year through the first grade. For the children in the E group who had had three years of enrichment, there was a significant gain over the C group of 8 months in reading achievement by the end of first grade, a score above national norms. This result is in keeping with the general finding that enrichment shows a greater effect on scholastic achievement than on IQ per se.

Many studies have employed no control group selected on exactly the same basis as the experimental group. This makes it virtually impossible to evaluate the effect of the treatment on pre-test—post-test gain, and the problem is made more acute by the fact that enrichment studies often pick their subjects on the basis of their being below the average IQ of the population of disadvantaged children from which they are selected. This makes statistical regression a certainty—the group's mean will increase by an appreciable amount because of the imperfect correlation between test-retest scores over, say, a one-year interval. Since this correlation is known to be considerably lower in younger than in older children, there will be considerably greater "gain" due to regression for younger groups of children. The net results of selecting especially backward children on the basis of IQ is that a gain in IQ can be predicted which is not at all attributable to the educational treatment given to the children. Studies using control groups nearly always show this gain in the control group, and only by subtracting the control group's gain from the experimental group's gain can we evaluate the magnitude of the treatment effect. Only the gain over and above that attributable to regression really counts.

Still another factor is involved in the inverse relationship generally found between children's age and the size of IQ gains in an enrichment program. Each single item gotten right in a test like the Stanford-Binet adds increasingly smaller increments to the IQ as children get older. Each Stanford-Binet test item, for example, is worth two months of mental age. At four years of age getting just two additional items right will boost an IQ of 85 up to 93. The same absolute amount of improvement in test performance at 10 years of age would boost an IQ of 85 up to only 88. The typical range of gains found in preschool enrichment programs, in the age range of 4 to 6, are about what would be expected from passing an additional two to four items in the Stanford-Binet. This amount of gain should not be surprising on a test which, for this age range, consists of items rather similar to the materials and activities traditionally found in nursery schools —blocks, animal pictures, puzzles, bead stringing, copying drawings, and the like. I once visited an experimental preschool using the Stanford-Binet to assess pretest—post-test gains, in which some of the Stanford-Binet test materials were openly accessible to the children throughout their time in the school as part of the enrichment paraphernalia. Years ago Reymert and Hinton (1940) noted this "easy gain" in the IQs of culturally disadvantaged preschoolers on tests depending on specific information such as being able to name parts of the body and knowing names of familiar objects. Children who have not picked up this information at home get it quickly in nursery school and kindergarten.

In addition to these factors, something else operates to boost scores five to ten points from first to second test, provided the first test is really the first. When I worked in a psychological clinic, I had to give individual intelligence tests to a variety of children, a good many of whom came from an impoverished background. Usually I felt these children were really brighter than their IQ would indicate. They often appeared inhibited in their responsiveness in the testing situation on their first visit to my office, and when this was the case I usually had them come in on two to four different days for half-hour sessions with me in a "play therapy" room, in which we did nothing more than get better acquainted by playing ball, using finger paints, drawing on the blackboard, making things out of clay, and so forth. As soon as the child seemed to be completely at home in this setting, I would retest him on a parallel form of the Stanford-Binet. A boost in IQ of 8 to 10 points or so was the rule; it rarely failed, but neither was the gain very often much above this. So I am inclined to doubt that IQ gains up to this amount in young disadvantaged children have much of anything to do with changes in ability. They are largely a result simply of getting a more accurate IQ by testing under more optimal conditions. Part of creating more optimal conditions in the case of disadvantaged children consists of giving at least two tests, the first only for practice and for letting the child get to know the examiner. I would put very little confidence in a single test score, especially if it is the child's first test and more especially if the child is from a poor background and of a different race from the examiner. But I also believe it is possible to obtain accurate assessments of a child's ability, and I would urge that attempts to evaluate preschool enrichment programs measure the gains against initially valid scores. If there is not evidence that this precaution has been taken, and if there is no control group, one might as well subtract at least 5 points from the gain scores as having little or nothing to do with real intellectual growth.

It is interesting that the IQ gains typically found in enrichment programs are of about the same magnitude and durability as those found in studies of the effects of direct coaching and practice on intelligence tests. The average IQ gain in such studies is about nine or ten points (Vernon, 1954).

What Is Really Changed When We Boost IQ? Test scores may increase after special educational treatment, but one must then ask which components of test variance account for the gain. Is it *g* that gains, or is it something less central to our concept of intelligence? We will not know for sure until someone does a factor analysis of pre- and post-test scores, including a number of "reference" tests that were not a part of the pre-test battery. We should also factor analyze the tests at the item level, to see which types of test items reflect the most gain. Are they the items with the highest cultural loadings? It is worth noting that the studies showing authentic gains used tests which are relatively high in cultural loading. I have found no studies that demonstrated gains in relatively noncultural or nonverbal tests like Cattell's Culture Fair Tests and Raven's Progressive Matrices.

Furthermore, if gain consists of actual improvement in cognitive skills rather than of acquisition of simple information, it must be asked whether the gain in skill represents the intellectual skill that the test normally measures, and which, because of the test's high heritability, presumably reflects some important, biologically based aspect of mental development. Let me cite one example. In a well-known experiment Gates and Taylor (1925) gave young children daily practice over several months in repeating auditory digit series, just like the digit span subtests in the Wechsler and Stanford-Binet. The practice resulted in a marked gain in the children's digit span, equivalent to an IQ gain of about 20 points. But when the children were retested after an interval of six months without practicing digit recall, their digit performance was precisely at the level expected for their mental age as determined by other tests. The gains had been lost, and the digit test once again accurately reflected the children's overall level of mental development, as it did before the practice period. The well-known later "fading" of IQ gains acquired early in enrichment programs may be a similar phenomenon.

But there is another phenomenon that probably is even more important as one of the factors working against the persistence of initial gains. This is the so-called "cumulative deficit" phenomenon, the fact that many children called disadvantaged show a decline in IQ from preschool age through at least elementary school. The term "cumulative deficit" may not be inappropriate in its connotations with respect to scholastic attainment, but it is probably a misleading misnomer when applied to the normal negatively accelerated growth rate of developmental characteristics such as intelligence. The same phenomenon can be seen in growth curves of stature, but no one would refer to the fact that some children gain height at a slower rate and level off at a lower asymptote as a "cumulative deficit." In short, it seems likely that some of the loss in initial gains is due to the more negatively accelerated growth curve for intelligence in disadvantaged children and is not necessarily due to waning or discontinuance of the instructional effort. The effort required to boost IQ from 80 to 90 at 4 or 5 years of age is miniscule compared to the effort that would be required by age 9 or 10. "Gains" for experimental children in this range, in fact, take the form of superiority over a control group which has declined in IQ; the "enriched" group is simply prevented from falling behind, so there is no absolute gain in IQ, but only an advantage relative to a declining control group. Because of the apparently ephemeral nature of the initial gains seen in preschool programs, judgments of these programs' effectiveness in making a significant impact on intellectual development should be based on long range results.

A further step in proving the effectiveness of a particular program is to demonstrate that it can be applied with comparable success by other individuals in other schools, and, if it is to be practicable on a large scale, to determine if it works in the hands of somewhat less inspired and less dedicated practitioners than the few who originated it or first put it into practice on a small scale. As an example of what can happen when a small-scale project gets translated to a large-scale one, we can note Kenneth B. Clark's (1963, p. 160) enthusiastic and optimistic description of a "total push" intensive compensatory program which originated in one school serving disadvantaged children in New York City, with initially encouraging results. Clark said, "These positive results can be duplicated in every school of this type." In fact, it was tried in 40 other New York schools, and became known as the Higher Horizons program. After three years of the program the children in it showed no gains whatever and even averaged slightly lower in achievement and IQ than similar children in ordinary schools (U.S. Commission on Civil Rights, 1967, p. 125).

Finally, little is known about the range of IQ most likely to show genuine gains under enrichment. None of the data I have seen in this area permits any clear judgment on this matter. It would be unwarranted to assume at this time that special educational programs push the whole IQ distribution up the scale, so that, for example, they would yield a higher precentage of children with IQs higher than two standard deviations above the mean. After a "total push" program, IQs, if they change at all, may no longer be normally distributed, so that the gains would not much affect the frequencies at the tails of the distribution. We simply do not know the answer to this at present, since the relevant data are lacking.

Hothouse or Fertilizer? There seems to be little doubt that a deprived environment can stunt intellectual development and that immersion in a good environment in early childhood can largely overcome the effects of deprivation, permitting the individual's genetic potential to be reflected in his performance. But can special enrichment and instructional procedures go beyond the prevention or amelioration of stunting? As Vandenberg (1968, p. 49) has asked, does enrichment act in a manner similar to a *hothouse,* forcing an early bloom which is nevertheless no different from a normal bloom, or does it act more like a *fertilizer*, producing bigger and better yields? There can be little question about the hothouse aspect of early stimulation and instruction. Within limits, children can learn many things at an earlier age than that at which they are normally taught in school. This is especially true of forms of associative learning which are mainly a function of time spent in the learning activity rather than of the development of more complex cognitive structures. While most children, for example, do not learn the alphabet until 5 or 6 years of age, they are fully capable of doing so at about 3, but it simply requires more time spent in learning. The cognitive structures involved are relatively simple as compared with, say, learning to copy a triangle or a diamond. Teaching a 3-year-old to copy a diamond is practically impossible; at five it is extremely difficult; at seven the child apparently needs no "teaching"—he copies the diamond easily. And the child of five who has been *taught* to copy the diamond seems to have learned something different from what the seven-year-old "knows" who can do it without being "taught." Though the final performance of the five-year-old and the seven-year-old may *look* alike, we know that the cognitive structures underlying their performance are different. Certain basic skills can be acquired either associatively by rote learning or cognitively by conceptual learning, and what superficially may appear to be the same performance may be acquired in preschoolers at an associative level, while at a conceptual level in older children. Both the four-year-old and the six-year-old may know that $2 + 2 = 4$, but this knowledge can be associative or cognitive. Insufficient attention has been given in preschool programs so far to the shift from associative to cognitive learning. The preschooler's capacity for associative learning is already quite well developed, but his cognitive or conceptual capacities are as yet rudimentary and will undergo their period of most rapid change between about five and seven years of age (White, 1965). We need to know more about what children can learn before age five that will transfer positively to later learning. Does learning something on an associative level facilitate or hinder learning the same content on a conceptual level?

While some preschool and compensatory programs have demonstrated earlier than normal learning of certain skills, the evidence for accelerating cognitive development or the speed of learning is practically nil. But usually this distinction is not made between sheer performance and the nature of the cognitive structures which support the gains in performance, and so the research leaves the issue in doubt. The answer to such questions is to be found in the study of the kinds and amount of transfer that result from some specific learning. The capacity for transfer of training is one of the essential aspects of what we mean by intelligence. The IQ gains reported in enrichment studies appear to be gains more in what Cattell calls 'crystallized," in contrast to "fluid," intelligence. This is not to say that gains of this type are not highly worthwhile. But having a clearer conception of just what the gains consist of will give us a better idea of how they can be most effectively followed up and of what can be expected of their effects on later learning and achievement.

Specific Programs. Hodges and Spicker (1967) have summarized a number of the more substantial preschool intervention studies designed to improve the intellectual capabilities and scholastic success of disadvantaged children. Here are some typical examples.

The *Indiana Project* focused on deprived Appalachian white children five years of age, with IQs in the range of 50 to 85. The children spent one year in a special kindergarten with a structured program designed to remedy specific diagnosed deficiencies of individual children in the areas of language development, fine motor coordination, concept formation, and socialization. Evaluation extended over two years, and gains were measured against three control groups: regular kindergarten, children who stayed at home during the kindergarten year, and children at home in another similar community. The average gain (measured against all three controls) after two years was 10.8 IQ points on the Stanford-Binet (final IQ 97.4) and 4.0 IQ points on the Peabody Picture Vocabulary Test (final IQ 90.4).

The *Perry Preschool Project* at Ypsilanti, Michigan, also was directed at disadvantaged preschool children with IQs between 50 and 85. The program was aimed at remedying lacks largely in the verbal prerequisites for first-grade learning and involved the parents as well as the children. There was a significant gain of 8.9 IQ points in the Stanford-Binet after one year of the preschool, but by the end of second grade the experimental group exceeded the controls, who had had no preschool attendance, by only 1.6 IQ points, a nonsignificant gain.

The *Early Training Project* under the direction of Gray and Klaus at Peabody College is described as a multiple intervention program, meaning that it included not only preschool enrichment but work with the disadvantaged children's mothers to increase their ability to stimulate their child's cognitive development at home. Two experimental groups, with two and three summers of preschool enrichment experience in a special school plus home visits by the training staff, experienced an average gain, four years after the start of the program, of 7.2 IQ points over a control group on the Stanford-Binet (final IQ of *E* group was 93.6).

The *Durham Education Improvement Program* (1966-1967b) has focused on preschool children from impoverished homes. The basic assumption of the program is stated as follows: "First, Durham's disadvantaged youngsters are considered normal at birth and potentially normal academic achievers, though they are frequently subjected to conditions jeopardizing their physical and emotional health. It is further assumed that they adapt to their environment according to the same laws of learning which apply to all children." The program is one of the most comprehensive and intensive efforts yet made to improve the educability of children from backgrounds of poverty. The IQ gains over about an eight to nine months' interval for various groups of preschoolers in the program are raw pre-post test gains, not gains over a control group. The average IQ gains on three different tests were 5.32 (Peabody Picture Vocabulary), 2.62 (Stanford-Binet), and 9.27 (Wechsler Intelligence Scale for children). In most cases, IQs changed from the 80s to the 90s.

The well-known Bereiter-Engelmann (1966) program at the University of Illinois is probably the most sharply focused of all. It aims not at all-round enrichment of the child's experience but at teaching specific cognitive skills, particularly of a logical, semantic nature (as contrasted with more diffuse "verbal stimulation"). The emphasis is on information processing skills considered essential for school learning. The Bereiter-Engelmann preschool is said to be academically oriented, since each day throughout the school year the children receive twenty-minute periods of intensive instruction in three major content areas—language, reading, and arithmetic. The instruction, in small groups, explicitly in-volves maintaining a high level of attention, motivation, and participation from every child. Overt and emphatic repetition by the children are important ingre-dients of the instructional process. The pre-post gains (not measured against a control group) in Stanford-Binet IQ over an eighteen months' period are about 8 to 10 points. Larger gains are shown in tests that have clearly identifiable con-tent which can reflect the areas receiving specific instruction, such as the Illinois Test of Psycholinguistic Abilities and tests of reading and arithmetic (Bereiter & Engelmann, 1968). The authors note that the gains are shared about equally by all children.

Bereiter and Engelmann, correctly, I believe, put less stock in the IQ gains than in the gains in scholastic performance achieved by the children in their program. They comment that the children's IQs were still remarkably low for children who performed at the academic level actually attained in the program. Their scholastic performance was commensurate with that of children 10 or 20 points higher in IQ. Such is the advantage of highly focused training—it can significantly boost the basic skills that count most. Bereiter and Engelmann (1966, p. 54) comment, ". . . to have taught children in a two-hour period per day enough over a broad area to bring the average IQ up to 110 or 120 would have been an impossibility." An important point of the Bereiter-Engelmann program is that it shows that scholastic performance—the acquisition of the basic skills—can be boosted much more, at least in the early years, than can the IQ, and that highly concentrated, direct instruction is more effective than more diffuse cultural en-richment.

The largest IQ gains I have seen and for which I was also able to examine the data and statistical analyses were reported by Karnes (1968), whose preschool program at the University of Illinois is based on an intensive attempt to amelio-rate specific learning deficits in disadvantaged three-year-old children. Between the average age of 3 years 3 months and 4 years 1 month, children in the program showed a gain of 16.9 points in the Stanford-Binet IQ, while a control group showed a loss of 2.8 over the same period, making for a net gain of 19.7 IQ points for the experimental group. Despite rather small samples (E = 15, C = 14), this gain is highly significant statistically (a probability of less than 1 in 1000 of occur-ring by chance). Even so, I believe such findings need to be replicated for proper evaluation, and the durability of the gains needs to be assessed by follow-up studies over several years. There remains the question of the extent to which speci-fic learning at age three affects cognitive structures which normally do not emerge until six or seven years of age and whether induced gains at an early level of mental development show appreciable "transfer" to later stages. It is hoped that investigators can keep sufficient track of children in preschool programs to permit a later follow-up which could answer these questions. An initial small sample size mitigates against this possibility, and so proper research programs should be planned accordingly.

"Expectancy Gain." Do disadvantaged children perform relatively poorly on intelligence tests because their teachers have low expectations for their ability? This belief has gained popular currency through an experiment by Rosenthal and Jacobson (1968). Their notion is that the teacher's expectations for the child's performance act as a self-fulfilling prophecy. Consequently, according to this hypothesis, one way to boost these children's intelligence, and presumably their general scholastic performance as well, is to cause teachers to hold out higher expectations of these children's ability. To test this idea, Rosenthal and Jacobson picked about five children at random from each of the classes in an elementary school and then informed the classroom teachers that, according to test results, the selected children were expected to show unusual intellectual gains in the coming year. Since the "high expectancy" children in each class were actually selected at random, the only way they differed from their classmates was presumably in the minds of their teachers. Group IQ tests administered by the teachers on three occasions during the school year showed a significantly larger gain in the "high expectancy" children than in their classmates. Both groups gained in IQ by amounts that are typically found as a result of direct coaching or of "total push" educational programs. Yet the authors note that "Nothing was done directly for the disadvantaged child at Oak School. There was no crash program to improve his reading ability, no special lesson plans, no extra time for tutoring, no trips to museums or art galleries. There was only the belief that the children bore watching, that they had intellectual competencies that would in due course be revealed" (p. 181). The net total IQ gain (i.e., Expectancy group minus Control group) for all grades was 3.8 points. Net gain in verbal IQ was 2.1; for Reasoning (nonverbal) IQ the gain was 7.2. Differences were largest in grades 1 and 2 and became negligible in higher grades. The statistical significance of the gains is open to question and permits no clear-cut conclusion. (The estimation of the error variance is at issue: the investigators emphasized the individual pupil's scores as the unit of analysis rather than the means of the E and C groups for each classroom as the unit. The latter procedure, which is regarded as more rigorous by many statisticians, yields statistically negligible results.)

Because of the questionable statistical significance of the results of this study, there may actually be no phenomenon that needs to be explained. Other questionable aspects of the conduct of the experiment make it mandatory that its results be replicated under better conditions before any conclusions from the study be taken seriously or used as a basis for educational policy. For example, the same form of the group-administered IQ test was used for each testing, so that specific practice gains were maximized. The teachers themselves administered the tests, which is a faux pas par excellence in research of this type. The dependability of teacher-administered group tests leaves much to be desired. Would any gains beyond those normally expected from general test familiarity have been found if the children's IQs had been accurately measured in the first place by individual tests administered by qualified psychometrists without knowledge of the purpose of the experiment? These are some of the conditions under which such an experiment must be conducted if it is to inspire any confidence in its results.

Conclusions About IQ Gains. The evidence so far suggests the tentative conclusion that the pay-off of preschool and compensatory programs in terms of IQ gains is small. Greater gains are possible in scholastic performance when instructional techniques are intensive and highly focused, as in the Bereiter-Engelmann program. Educators would probably do better to concern themselves with teaching basic skills directly than with attempting to boost overall cognitive development. By the same token, they should deemphasize IQ tests as a means of assessing gains, and use mainly direct tests of the skills the instructional program is intended to inculcate. The techniques for raising intelligence per se, in the sense of *g*, probably lie more in the province of the biological sciences than in psychology and education.

Gordon and Wilkerson (1966, pp. 158-159) have made what seems to me perhaps the wisest statement I have encountered regarding the proper aims of intervention programs:

. . . the unexpressed purpose of most compensatory programs is to make disadvantaged children as much as possible like the kinds of children with whom the school has been successful, and our standard of educational success is how well they approximate middle-class children in school performance. It is not at all clear that the concept of compensatory education is the one which will most appropriately meet the problems of the disadvantaged. These children are *not* middle-class children, many of them never *will* be, and they can never be anything but second-rate as long as they are thought of as potentially middle-class children.... At best they are different, and an approach which views this difference merely as something to be overcome is probably doomed to failure.

"Learning Quotient" versus Intelligence Quotient

If many of the children called culturally disadvantaged are indeed "different" in ways that have educational implications, we must learn as much as possible about the real nature of these differences. To what extent do the differences consist of more than just the well-known differences in IQ and scholastic achievement, and, of course, the obvious differences in cultural advantages in the home?

Evidence is now emerging that there are stable ethnic differences in *patterns* of ability and that these patterns are invariant across wide socioeconomic differences (Lesser, Fifer, & Clark, 1965; Stodolsky & Lesser, 1967). Middle-class and lower-class groups differed about one standard deviation on all four abilities (Verbal, Reasoning, Number, Space) measured by Lesser and his co-workers, but the profile or pattern of scores was distinctively different for Chinese, Jewish, Negro, and Puerto Rican children, regardless of their social class. Such differences in patterns of ability are bound to interact with school instruction. The important question is how many other abilities there are that are not tapped by conventional tests for which there exist individual and group differences that interact with methods of instruction.

Through our research in Berkeley we are beginning to perceive what seems to be a very significant set of relationships with respect to patterns of ability which, unlike those of Lesser et al., seem to interact more with social class than with ethnic background.

In brief, we are finding that a unidimensional concept of intelligence is quite inadequate as a basis for understanding social class differences in ability. For example, the magnitude of test score differences between lower- and middle-class children does not always correspond to the apparent "cultural loading" of the test. Some of the least culturally loaded tests show the largest differences between lower- and middle-class children. At least two dimensions must be postulated to comprehend the SES differences reported in the literature and found in our laboratory (see Jensen, 1968c, 1968d). These two dimensions and the hypothetical location of various test loadings on each dimension are shown in Figure 17. The horizontal axis represents the degree of cultural loading of the test. It is defined by the test's heritability. I have argued elsewhere (Jensen, 1968c) that the heritability index for a test is probably our best objective criterion of its culture-

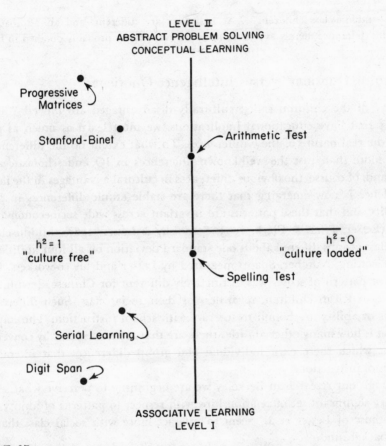

LEVEL II
ABSTRACT PROBLEM SOLVING
CONCEPTUAL LEARNING

Progressive
Matrices

Stanford-Binet

Arithmetic Test

$h^2 = 1$
"culture free"

$h^2 = O$
"culture loaded"

Spelling Test

Serial Learning

Digit Span

ASSOCIATIVE LEARNING
LEVEL I

FIGURE 17.

The two-dimensional space required for comprehending social class differences in performance on tests of intelligence, learning ability, and scholastic achievement. The locations of the various "tests" are hypothetical.

fairness. Just because tests do not stand at one or the other extreme of this continuum does not mean that the concept of culture-fairness is not useful in discussing psychological tests. The vertical axis in Figure 17 represents a continuum ranging from "simple" associative learning to complex cognitive or conceptual learning. I have hypothesized two genotypically distinct basic processes underlying this continuum, labeled Level I (associative ability) and Level II (conceptual ability). Level I involves the neural registration and consolidation of stimulus inputs and the formation of associations. There is relatively little transformation of the input, so there is a high correspondence between the forms of the stimulus input and the form of the response output. Level I ability is tapped mostly by tests such as digit memory, serial rote learning, selective trial-and-error learning with reinforcement (feedback) for correct responses, and in slightly less "pure" form by free recall of visually or verbally presented materials, and paired-associate learning. Level II abilities, on the other hand, involve self-initiated elaboration and transformation of the stimulus input before it eventuates in an overt response. Concept learning and problem solving are good examples. The subject must actively manipulate the input to arrive at the output. This ability is best measured by intelligence tests with a low cultural loading and a high loading on g—for example, Raven's Progressive Matrices.

Social class differences in test performance are more strongly associated with the vertical dimension in Figure 17 than with the horizontal.

Associative Learning Ability

Teachers of the disadvantaged have often remarked that many of these children seem much brighter than their IQs would lead one to expect, and that, even though their scholastic performance is usually as poor as that of middle-class children of similar IQ, the disadvantaged children usually appear much brighter in nonscholastic ways than do their middle-class counterparts in IQ. A lower-class child coming into a new class, for example, will learn the names of 20 or 30 children in a few days, will quickly pick up the rules and the know-how of various games on the playground, and so on—a kind of performance that would seem to belie his IQ, which may even be as low as 60. This gives the impression that the test is "unfair" to the disadvantaged child, since middle-class children in this range of IQ will spend a year in a classroom without learning the names of more than a few classmates, and they seem almost as inept on the playground and in social interaction as they are in their academic work.

We have objectified this observation by devising tests which can reveal these differences. The tests measure associative learning ability and show how fast a child can learn something relatively new and unfamiliar, right in the test situation. The child's performance does not depend primarily, as it would in conventional IQ tests, upon what he has already learned at home or elsewhere before he comes to take the test. We simply give him something to learn, under conditions which permit us to measure the rate and thoroughness of the learning. The tasks most frequently used are various forms of auditory digit memory, learning the serial order of a number of familiar objects or pictures of objects, learning to associate pairs of pictures of familiar objects, and free recall of names or objects presented from one to five times in a random order.

Our findings with these tests, which have been presented in greater detail elsewhere (Jensen, 1968a, 1968b, 1968d, 1968e; Jensen, 1968f; Jensen & Rohwer, 1968), seem to me to be of great potential importance to the education of many of the children called disadvantaged. What we are finding, briefly, is this: lower-class children, whether white, Negro, or Mexican-American, perform as well on these direct learning tests as do middle-class children. Lower-class children in the IQ range of about 60 to 80 do markedly *better* than middle-class children who are in this range of IQ. Above about IQ 100, on the other hand, there is little or no difference between social class groups on the learning tests.

At first we thought we had finally discovered a measure of "culture-fair" testing, since we found no significant SES differences on these learning tests. But we can no longer reconcile this interpretation with all the facts now available. Some of the low SES children with low IQs on culturally loaded tests, like the Peabody Picture Vocabulary Tests, do very well on our learning tests, but do not have higher IQs on less culturally loaded tests of g, like the Progressive Matrices. It appears that we are dealing here with two kinds of abilities—associative learning ability (Level I) and cognitive or conceptual learning and problem-solving ability (Level II).

One particular test—free recall—shows the distinction quite well, since a slight variation in the test procedure makes the difference between whether it measures Level I or Level II. This is important, because it is sometimes claimed that low SES children do better on our learning tests than on IQ tests because the former are more interesting or more "relevant" to them, and thus make them more highly motivated to perform at their best. This is not a valid interpretation, since when essentially the same task is made either "associative" or "cognitive," we get differences of about one standard deviation in the mean scores of lower- and middle-class children. For example, 20 unrelated familiar objects (doll, toy car, comb, cup, etc.) are shown to children who are then asked to recall as many objects as they can in any order that may come to mind. The random presentation and recall are repeated five times to obtain a more reliable score. Lower- and middle-class elementary school children perform about the same on this task, although they differ some 15 to 20 points in IQ. This free recall test has a low correlation with IQ and the correlation is lower for the low SES children. But then we can change the recall test so that it gives quite different results.

This is shown in an experiment from our laboratory by Glasman (1968). (In this study SES and race are confounded, since the low SES group were Negro children and the middle SES group were white.) Again, 20 familiar objects are presented, but this time the objects are selected so that they can be classified into one of four categories, *animals, furniture, clothing,* or *foods.* There are five items in each of the four categories, but all 20 items are presented in a random order on each trial. Under this condition a large social class difference shows up: the low SES children perform only slightly better on the average than they did on the uncategorized objects, while the middle SES children show a great improvement in performance which puts their scores about one standard deviation above the low SES children. Furthermore, there is much greater evidence of "clustering" the items in free recall for the middle SES than for the low SES children. That is, the middle-class children rearrange the input in such a way that the order of output in recall corresponds to the categories to which the objects may be assigned. The low SES children show less clustering in this fashion, although many show rather idiosyncratic pair-wise "clusters" that persist from trial to trial. There is a high correlation between the strength of the clustering tendency and the amount of recall. Also, clustering tendency is strongly related to age. Kindergarteners, for example, show little difference between recall of categorized and uncategorized lists, and at this age SES differences in performance are nil. By fourth or fifth grade, however, the SES differences in clustering tendency are great, with a correspondingly large difference in ability to recall categorized lists.

It is interesting, also, that the recall of categorized lists correlates highly with IQ. In fact, when mental age or IQ is partialled out of the results, there are no significant remaining SES differences in recall. Post-test interviews showed that the recall differences for the two social class groups cannot be attributed to the low SES group's not knowing the category names. The children know the categories but tend not to use them spontaneously in recalling the list.

In general, we find that Level I associative learning tasks correlate very substantially with IQ among middle-class children but have very low correlations with IQ among lower-class children (Jensen, 1968b). The reason for this difference in correlations can be traced back to the form of the scatter diagrams for the middle and low SES groups, which is shown schematically in Figure 18. Since large representative samples of the entire school population have not been studied so far, the exact form of the correlation scatter diagram has not yet been well established, but the schematic portrayal of Figure 18 is what could be most reasonably hypothesized on the basis of several lines of evidence now available. (Data

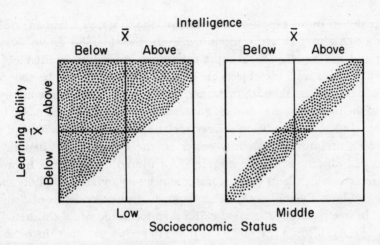

FIGURE 18.

Schematic illustration of the essential form of the correlation scatter-diagram for the relationship between associative learning ability and IQ in Low SES and Upper-Middle SES groups.

on a representative sample of 5000 children given Level I and Level II tests are now being analyzed to establish the forms of the correlation plots for low and middle SES groups.) The form of the correlation as it now appears suggests a hierarchical arrangement of mental abilities, such that Level I ability is necessary but not sufficient for Level II. That is, high performance on Level II tasks depends upon better than average ability on Level I, but the reverse does not hold. If this is true, the data can be understood in terms of one additional hypothesis, namely, that Level I ability is distributed about the same in all social class groups, while Level II ability is distributed differently in lower and middle SES groups. The hypothesis is expressed graphically in Figure 19. Heritability studies of Level II tests cause me to believe that Level II processes are not just the result of interaction between Level I learning ability and experientially acquired strategies or learning sets. That learning is necessary for Level II no one doubts, but certain neural structures must also be available for Level II abilities to develop, and these are conceived of as being different from the neural structures underlying Level I. The genetic factors involved in each of these types of ability are presumed to have become differentially distributed in the population as a function of social class, since Level II has been most important for scholastic performance under the traditional methods of instruction.

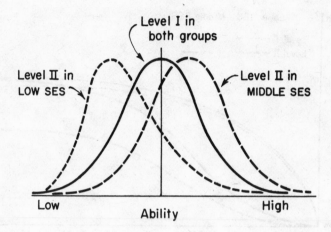

FIGURE 19.

Hypothetical distributions of Level I (solid line) and Level II (dashed line) abilities in middle-class and culturally disadvantaged populations.

From evidence on age differences in different tasks on the Level I—Level II continuum (e.g., Jensen & Rohwer, 1965), I have suggested one additional hypothesis concerning the developmental rates of Level I and Level II abilities in lower and middle SES groups, as depicted in Figure 20. Level I abilities are seen as developing rapidly and as having about the same course of development and final level in both lower and middle SES groups. Level II abilities, by contrast, develop slowly at first, attain prominence between four and six years of age, and show an increasing difference between the SES groups with increasing age. This formulation is consistent with the increasing SES differences in mental age on standard IQ tests, which tap mostly Level II ability.

Thus, ordinary IQ tests are not seen as being "unfair" in the sense of yielding inaccurate or invalid measures for the many disadvantaged children who obtain low scores. If they are unfair, it is because they tap only one part of the total spectrum of mental abilities and do not reveal that aspect of mental ability which may be the disadvantaged child's strongest point—the ability for associative learning.

Since traditional methods of classroom instruction were evolved in populations having a predominantly middle-class pattern of abilities, they put great emphasis

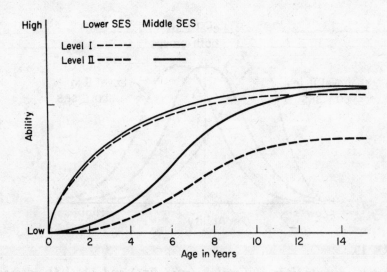

FIGURE 20.

Hypothetical growth curves for Level I and Level II abilities in middle SES and low SES populations.

on cognitive learning rather than associative learning. And in the post-Sputnik era, education has seen an increased emphasis on cognitive and conceptual learning, much to the disadvantage of many children whose mode of learning is predominantly associative. Many of the basic skills can be learned by various means, and an educational system that puts inordinate emphasis on only one mode or style of learning will obtain meager results from the children who do not fit this pattern. At present, I believe that the educational system—even as it falteringly attempts to help the disadvantaged—operates in such a way as to maximize the importance of Level II (i.e., intelligence or g) as a source of variance in scholastic performance. Too often, if a child does not learn the school subject matter when taught in a way that depends largely on being average or above average on g, he does not learn at all, so that we find high school students who have failed to learn basic skills which they could easily have learned many years earlier by means that do not depend much on g. It may well be true that many children today are confronted in our schools with an educational philosophy and methodology which were mainly shaped in the past, entirely without any roots in these children's genetic and cultural heritage. The educational system was never

116

allowed to evolve in such a way as to maximize the actual potential for learning that is latent in these children's patterns of abilities. If a child cannot show that he "understands" the meaning of $1 + 1 = 2$ in some abstract, verbal, cognitive sense, he is, in effect, not allowed to go on to learn $2 + 2 = 4$. I am reasonably convinced that all the basic scholastic skills can be learned by children with normal Level I learning ability, provided the instructional techniques do not make g (i.e., Level II) the *sine qua non* of being able to learn. Educational researchers must discover and devise teaching methods that capitalize on existing abilities for the acquisition of those basic skills which students will need in order to get good jobs when they leave school. I believe there will be greater rewards for all concerned if we further explore different types of abilities and modes of learning, and seek to discover how these various abilities can serve the aims of education. This seems more promising than acting as though only one pattern of abilities, emphasizing g, can succeed educationally, and therefore trying to inculcate this one ability pattern in all children.

If the theories I have briefly outlined here become fully substantiated, the next step will be to develop the techniques by which school learning can be most effectively achieved in accordance with different patterns of ability. By all means, schools must discover g wherever it exists and see to it that its educational correlates are fully encouraged and cultivated. There can be little doubt that certain educational and occupational attainments depend more upon g than upon any other single ability. But schools must also be able to find ways of utilizing other strengths in children whose major strength is not of the cognitive variety. One of the great and relatively untapped reservoirs of mental ability in the disadvantaged, it appears from our research, is the basic ability to learn. We can do more to marshal this strength for educational purposes.

If diversity of mental abilities, as of most other human characteristics, is a basic fact of nature, as the evidence indicates, and if the ideal of universal education is to be successfully pursued, it seems a reasonable conclusion that schools and society must provide a range and diversity of educational methods, programs, and goals, and of occupational opportunities, just as wide as the range of human abilities. Accordingly, the ideal of equality of educational opportunity should not be interpreted as uniformity of facilities, instructional techniques, and educational aims for all children. Diversity rather than uniformity of approaches and aims would seem to be the key to making education rewarding for children of different patterns of ability. The reality of individual differences thus need not mean educational rewards for some children and frustration and defeat for others.

References

Altus, W. D. Birth order and its sequelae. *Science,* 1966, 151, 44-59.

Anastasi, A. Intelligence and family size. *Psychol. Bull.,* 1956, 53, 187-209.

Bajema, C. J. Estimation of the direction and intensity of natural selection in relation to human intelligence by means of the intrinsic rate of natural increase. *Eugen. Quart.,* 1963, 10, 175-187.

Bajema, C. J. Relation of fertility to educational attainment in a Kalamazoo public school population: A follow-up study. *Eugen. Quart.,* 1966, 13, 306-315.

Bayley, N. Research in child development: A longitudinal perspective. *Merrill-Palmer Quart. Behav. Developm.,* 1965, 11, 183-208. (b)

Bayley, N. Comparisons of mental and motor test scores for ages 1-15 months by sex, birth order, race, geographical location, and education of parents. *Child Developm.,* 1965, 36, 379-411. (a)

Bayley, N. Learning in adulthood: The role of intelligence. In H. J. Klausmeier & C. W. Harris (Eds.), *Analyses of concept learning.* New York: Academic Press, 1966. Pp. 117-138.

Bayley, N. Behavioral correlates of mental growth: Birth to thirty-six years. *Amer. Psychol.,* 1968, 23, 1-17.

Bereiter, C., & Engelmann, S. *Teaching disadvantaged children in the preschool.* Englewood Cliffs, N.J.: Prentice-Hall, 1966.

Bereiter, C., & Engelmann, S. An academically oriented preschool for disadvantaged children: Results from the initial experimental group. In D. W. Brison & J. Hill (Eds.), *Psychology and early childhood education.* Ontario Institute for Studies in Education, 1968. No. 4. Pp. 17-36.

Bilodeau, E. A. (Ed.), *Acquisition of skill.* New York: Academic Press, 1966.

Brison, D. W. Can and should learning be accelerated? In D. W. Brison (Ed.), *Accelerated learning and fostering creativity.* Toronto, Canada: Ontario Institute for Studies in Education, 1968. Pp. 5-9.

Bronfenbrenner, U. The psychological costs of quality and equality in education. *Child Developm.,* 1967, 38, 909-925.

Buck, C. Discussion of "Culturally related reproductive factors in mental retardation" by Graves et al. Paper read at Conference on Sociocultural Aspects of Mental Retardation, Peabody College, Nashville, Tenn., June, 1968.

Burks, B. S. The relative influence of nature and nurture upon mental development: A comparative study of foster parent-foster child resemblance and true parent-true child resemblance. *Yearb. Nat. Soc. Stud. Educ.,* 1928, 27, (I), 219-316.

Burt, C. The evidence for the concept of intelligence. *Brit. J. educ. Psychol.,* 1955, 25, 158-177.

Burt, C. The distribution of intelligence. *Brit. J. Psychol.,* 1957, 48, 161-175.

Burt, C. The inheritance of mental ability. *Amer. Psychol.,* 1958, 13, 1-15.

Burt, C. Class difference in general intelligence: III. *Brit. J. Stat. Psychol.,* 1959, 12, 15-33.

Burt, C. Intelligence and social mobility. *Brit. J. Stat. Psychol.,* 1961, 14, 3-24.

Burt, C. Is intelligence distributed normally? *Brit. J. Stat. Psychol.,* 1963, 16, 175-190.

Burt, C. The genetic determination of differences in intelligence: A study of monozygotic twins reared together and apart. *Brit. J. Psychol.,* 1966, 57, 137-153.

Burt, C. Mental capacity and its critics. *Bull. Brit. Psychol. Soc.,* 1968, 21, 11-18.

Burt, C., & Howard, M. The multifactorial theory of inheritance and its application to intelligence. *Brit. J. Stat. Psychol.,* 1956, 9, 95-131.

Burt, C., & Howard, M. The relative influence of heredity and environment on assessments of intelligence. *Brit. J. Stat. Psychol.,* 1957, 10, 99-104.

Bloom, B. S. *Stability and change in human characteristics.* New York: Wiley, 1964.

Carter, C. O. Differential fertility by intelligence. In J. E. Meade & A. S. Parkes (Eds.), *Genetic and environmental factors in human ability.* New York: Plenum Press, 1966. Pp. 185-200.

Cattell, R. B. The multiple abstract variance analysis equations and solutions: For nature-nurture research on continuous variables. *Psychol. Rev.,* 1960, 67, 353-372.

Cattell, R. B. Theory of fluid and crystallized intelligence: A critical experiment. *J. educ. Psychol.*, 1963, 54, 1-22.

Churchill, J. A., Neff, J. W., & Caldwell, D. F. Birth weight and intelligence. *Obstetrics and gynecology*, 1966, 28, 425-429.

Clark, K. B. Educational stimulation of racially disadvantaged children. In A. H. Passow (Ed.), *Education in depressed areas*. New York: Teachers College Press, Columbia Univer., 1963. Pp. 142-162.

Coleman, J. S., et al. *Equality of educational opportunity*. U.S. Dept. of Health, Education, & Welfare, 1966.

Cooper, R., & Zubek, J. Effects of enriched and restricted early environments on the learning ability of bright and dull rats. *Canad. J. Psychol.*, 1958, 12, 159-164.

Cravioto, J. Malnutrition and behavioral development in the preschool child. *Pre-school child malnutrition*. National Health Science, Public., 1966, No. 1282.

Cravioto, J., De Licardie, E. R., & Birch, H. G. Nutrition, growth, and neurointegrative development: An experimental and ecologic study. *Pediatrics*, 1966, 38, 319-372.

Davis, K. Final note on a case of extreme isolation. *Amer. J. Sociol.*, 1947, 57, 432-457.

Deutsch, M., Katz, I., & Jensen, A. R. (Eds.) *Social class, race, and psychological development*. New York: Holt, Rinehart & Winston, 1968.

Dobzhansky, T. Genetic differences between people cannot be ignored. *Scientific Res.*, 1968, 3, 32-33. (a)

Dobzhansky, T. On genetics, sociology, and politics. *Perspect. Biol. Med.*, 1968, 11, 544-554. (b)

Dreger, R. M., & Miller, K. S. Comparative psychological studies of Negroes and whites in the United States. *Psychol. Bull.*, 1960, 57, 361-402.

Dreger, R. M., & Miller, K. S. Comparative psychological studies of Negroes and whites in the United States: 1959-1965. *Psychol. Bull.*, 1968 (*Monogr. Suppl.* 70, No. 3, Part 2).

Duncan, O. D. Is the intelligence of the general population declining? *Amer. sociol. Rev.*, 1952, 17, 401-407.

Duncan, O. D., Featherman, D. L., & Duncan, B. Socioeconomic background and occupational achievement: Extensions of a basic model. Final Report, Project No. 5-0074 (EO-191) U.S. Dept. of Health, Education, and Welfare, Office of Education, Bureau of Research, May, 1968.

Durham Education Improvement Program, 1966-1967. (a)

Durham Education Improvement Program, Research, 1966-1967. (b)

Dustman, R. E., & Beck, E. C. The visually evoked potential in twins. *Electroenceph. clin. Neurophysiol.*, 1965, 19, 570-575.

Eckland, B. K. Genetics and sociology: A reconsideration. *Amer. Soc. Rev.*, 1967, 32, 173-194.

Eells, K., et al. *Intelligence and cultural differences*. Chicago: Univ. Chicago Press, 1951.

Erlenmeyer-Kimling, L., & Jarvik, L. F. Genetics and intelligence: a review. *Science*, 1963, 142, 1477-1479.

Falconer, D. S. *An introduction to quantitative genetics*. New York: Ronald Press, 1960.

Freeman, R. A. Schools and the elusive 'average children' concept. *Wall Street Journal*, July 8, 1968, p. 12.

Fuller, J. L., & Thompson, W. R. *Behavior genetics*. New York: Wiley, 1960.

Gates, A. I., & Taylor, G. A. An experimental study of the nature of improvement resulting from practice in mental function. *J. educ. Psychol.*, 1925, 16, 583-593.

Geber, M. The psycho-motor development of African children in the first year, and the influence of maternal behavior. *J. soc. Psychol.*, 1958, 47, 185-195.

Geber, M., & Dean, R. F. A. The state of development of newborn African children. *Lancet*, 1957, 1216-1219.

Ghiselli, E. E. The measurement of occupational aptitude. *Univer. of California Publ. in Psychol.*, Vol. 8, No. 2. Berkeley, Calif.: Univer. of California Press, 1955.

Glasman, L. D. A social-class comparison of conceptual processes in children's free recall. Unpublished doctoral dissertation, Univer. of California, 1968.

Goertzel, V., & Goertzel, M. G. *Cradles of eminence*. London: Constable, 1962.

Goodenough, F. L. New evidence on environmental influence on intelligence. *Yearb. Nat. Soc. Stud. Educ.*, 1940, 39, Part I, 307-365.

Gordon, E. W., & Wilkerson, D. A. *Compensatory education for the disadvantaged.* New York: College Entrance Examination Board, 1966.

Gottesman, I. Genetic aspects of intelligent behavior. In N. R. Ellis (Ed.), *Handbook of mental deficiency.* New York: McGraw-Hill, 1963. Pp. 253-296.

Gottesman, I. Biogenetics of race and class. In M. Deutsch, I. Katz, & A. R. Jensen (Eds.), *Social class, race, and psychological development.* New York: Holt, Rinehart & Winston, 1968. Pp. 11-51.

Graves, W. L., Freeman, M. G., & Thompson, J. D. Culturally related reproductive factors in mental retardation. Paper read at Conference on Sociocultural Aspects of Mental Retardation, Peabody College, Nashville, Tenn., June, 1968.

Guilford, J. P. *The nature of human intelligence.* New York: McGraw-Hill, 1967.

Hardy, J. B. Perinatal factors and intelligence. In S. F. Osler & R. E. Cooke (Eds.), *The biosocial basis of mental retardation.* Baltimore, Md.: The Johns Hopkins Press, 1965. Pp. 35-60.

Harlow, H. F., & Griffin, G. Induced mental and social deficits in Rhesus monkeys. In S. F. Osler & R. E. Cooke (Eds.), *The biosocial basis of mental retardation.* Baltimore, Md.: The Johns Hopkins Press, 1965. Pp. 87-106.

Harlow, H. F., & Harlow, M. K. The mind of man. In *Yearbook of science and technology.* New York: McGraw-Hill, 1962.

Harrell, R. F., Woodyard, E., & Gates, A. I. *The effects of mothers' diets on the intelligence of offspring.* New York: Bureau of Publications, Teachers College, 1955.

Heber, R. Research on education and habilitation of the mentally retarded. Paper read at Conference on Sociocultural Aspects of Mental Retardation, Peabody College, Nashville, Tenn., June, 1968.

Heber, R., Dever, R., & Conry, J. The influence of environmental and genetic variables on intellectual development. In H. J. Prehm, L. A. Hamerlynck, & J. E. Crosson (Eds.), *Behavioral Research in mental retardation.* Eugene, Oregon: Univer. of Oregon Press, 1968. Pp. 1-23.

Heyns, O. S. *Abdominal decompression.* Johannesburg: Witwatersrand Univer. Press, 1963.

Higgins, J., Reed, S., & Reed, E. Intelligence and family size: A paradox resolved. *Eugen. Quart.,* 1962, 9, 84-90.

Hill, A. C., & Jaffee, F. S. Negro fertility and family size preferences. In T. Parsons & K. B. Clark (Eds.), *The Negro American.* Cambridge, Mass.: Houghton-Mifflin, 1966. Pp. 134-159.

Hodges, W. L., & Spicker, H. H. The effects of preschool experiences on culturally deprived children. In W. W. Hartup & N. L. Smothergill (Eds.), *The young child: Reviews oj research.* Washington, D.C.: National Association for the Education of Young Children, 1967. Pp. 262-289.

Honzik, M. P. Developmental studies of parent-child resemblance in intelligence. *Child Developm.,* 1957, 28, 215-228.

Honzik, M. P. The mental and motor test performances of infants diagnosed or suspected of brain injury. Final Rep., Contract SA 43 PH 2426. Washington, D.C.: National Institute of Health, National Institute of Neurological Diseases and Blindness, Collaborative Research, May, 1962.

Hunt, J. McV. *Intelligence and experience.* New York: Ronald Press, 1961.

Huntley, R. M. C. Heritability of intelligence. In J. E. Meade & A. S. Parker (Eds.), *Genetic and environmental factors in human ability.* New York: Plenum Press, 1966. Pp. 201-218.

Husén, T. Abilities of twins. *Scand. J. Psychol.,* 1960, 1, 125-135.

Jensen, A. R. Estimation of the limits of heritability of traits by comparison of monozygotic and dizygotic twins. *Proc. Nat. Acad. Sci.,* 1967, 58, 149-157. (a)

Jensen, A. R. Varieties of individual differences in learning. In Gagné, R. M. (Ed.), *Learning and individual differences.* Columbus, Ohio: Merrill, 1967. (b)

Jensen, A. R. Social class, race, and genetics: Implications for education. *Amer. Educ. Res. J.,* 1968, 5, 1-42. (a)

Jensen, A. R. Patterns of mental ability and socioeconomic status. *Proc. Nat. Acad. Sci.*, 1968, 60, 1330-1337. (b)

Jensen, A. R. Another look at culture-fair tests. In *Western Regional Conference on Testing Problems, Proceedings for 1968,* "Measurement for Educational Planning." Berkeley, Calif.: Educational Testing Service, Western Office, 1968. Pp. 50-104. (c)

Jensen, A. R. Intelligence, learning ability, and socioeconomic status. *J. spec. Educ.*, 1968. (d)

Jensen, A. R. Social class and verbal learning. In M. Deutsch, I. Katz, & A. R. Jensen (Eds.), *Social class, race, and psychological development.* New York: Holt, Rinehart & Winston, 1968. Pp. 115-174. (e)

Jensen, A. R. The culturally disadvantaged and the heredity-environment uncertainty. In J. Helmuth (Ed.), *The culturally disadvantaged child.* Vol. 2. Seattle, Wash.: Special Child Publications, 1968. (f)

Jensen, A. R., & Rohwer, W. D., Jr. Syntactical mediation of serial and paired-associate learning as a function of age. *Child Developm.*, 1965, 36, 601-608.

Jensen, A. R., & Rohwer, W. D., Jr. Mental retardation, mental age, and learning rate. *J. educ. Psychol.*, 1968.

Jepsen, N. P., & Bredmose, G. V. Investigations into the age of mentally deficient women at their first delivery. *Acta Psychiat. Scand.*, 1956 (Monogr. Suppl. 108), Pp. 203-210.

Jones, H. E. The environment and mental development. In L. Carmichael (Ed.), *Manual of child psychology.* (2nd ed.) New York: Wiley, 1954. Pp. 631-696.

Karnes, M. B. A research program to determine the effects of various preschool intervention programs on the development of disadvantaged children and the strategic age for such intervention. Paper read at Amer. Educ. Res. Ass., Chicago, Feb., 1968.

Kempthorne, O. *An introduction to genetic statistics.* New York: Wiley, 1957.

Kennedy, W. A., Van De Riet, V., & White, J. C., Jr. A normative sample of intelligence and achievement of Negro elementary school children in the Southeastern United States. *Monogr. Soc. Res. Child Developm.*, 1963, 28, No. 6.

Kushlick, A. Assessing the size of the problem of subnormality. In J. E. Meade & A. S. Parkes (Eds.), *Genetic and environmental factors in human ability.* New York: Plenum Press, 1966. Pp. 121-147.

Kuttner, R. E. *Biochemical anthropology.* In R. E. Kuttner (Ed.), *Race and modern science.* New York: Social Science Press, 1967. Pp. 197-222.

Kuttner, R. E. Letters to and from the editor. *Perspect. Biol. Med.*, 1968, 11, 707-709.

Lawrence, E. M. An investigation into the relation between intelligence and inheritance. *Brit. J. Psychol. Monogr. Suppl.*, 1931, 16, No. 5.

Leahy, A. M. Nature-nurture and intelligence. *Genet. Psychol. Monogr.*, 1935, 17, 241-305.

Lesser, G. S., Fifer, G., & Clark, D. H. Mental abilities of children from different social-class and cultural groups. *Monogr. Soc. Res. Child Developm.*, 1965, 30, (4).

Lindzey, G. Some remarks concerning incest, the incest taboo, and psychoanalytic theory. *Amer. Psychol.*, 1967, 22, 1051-1059.

Loehlin, J. C. Psychological genetics, from the study of human behavior. In R. B. Cattell (Ed.), *Handbook of modern personality theory.* New York: Aldine, in press.

Loevinger, J. On the proportional contributions of differences in nature and nurture to differences in intelligence. *Psychol. Bull.*, 1943, 40, 725-756.

Medical World News. Using speed of brain waves to test IQ. 1968, 9, 26.

Mitra, S. Income, socioeconomic status, and fertility in the United States. *Eugen. Quart.*, 1966, 13, 223-230.

Money, J. Two cytogenetic syndromes: Psychologic comparisons 1. Intelligence and specific-factor quotients. *J. psychiat. Res.*, 1964, 2, 223-231.

Moynihan, D. P. *The Negro family.* Washington, D.C.: Office of Policy Planning and Research, United States Department of Labor, 1965.

Moynihan, D. P. Employment, income, and the ordeal of the Negro family. In T. Parsons & K. B. Clark (Eds.), *The Negro American.* Cambridge, Mass.: Houghton-Mifflin, 1966. Pp. 134-159.

National Academy of Sciences. Racial studies: Academy states position on call for new research. *Science*, 1967, **158**, 892-893.

Naylor, A. E., & Myrianthopoulos, N. C. The relation of ethnic and selected socioeconomic factors to human birth-weight. *Ann. Hum. Genet.*, 1967, **31**, 71-83.

Nelson, G. K., & Dean, R. F. A. *Bull. World Health Organ.*, 1959, **21**, 779. Cited by G. Cravioto, Malnutrition and behavioral development in the preschool child. *Pre-school child malnutrition.* National Health Science, 1966, Public. No. 1282.

Newman, H. H., Freeman, F. N., & Holzinger, K. J. *Twins: A study of heredity and environment.* Chicago: Univ. of Chicago Press, 1937.

Nichols, R. C., & Bilbro, W. C., Jr. The diagnosis of twin zygosity. *Acta genet.*, 1966, **16**, 265-275.

Pettigrew, T. *A profile of the Negro American.* Princeton, N. J.: Van Nostrand, 1964.

Powledge, F. *To change a child—A report on the Institute for Developmental Studies.* Chicago: Quadrangle Books, 1967.

Reed, E. W., & Reed, S. C. *Mental retardation: A family study.* Philadelphia: W. B. Saunders Co., 1965.

Research Profile No. 11. Summary of progress in childhood disorders of the brain and nervous system. Washington, D.C.: Public Health Service, 1965.

Reymert, M. L., & Hinton, R. T., Jr. The effect of a change to a relatively superior environment upon the IQs of one hundred children. *Yearb. Nat. Soc. Stud. Educ.*, 1940, **39**, (I), 255-268.

Rimland, B. *Infantile autism.* New York: Appleton-Century-Crofts, 1964.

Roberts, J. A. F. The genetics of mental deficiency. *Eugen. Rev.*, 1952, **44**, 71-83.

Roberts, R. C. Some concepts and methods in quantitative genetics. In J. Hirsch (Ed.), *Behavior-genetic analysis.* New York: McGraw-Hill, 1967. Pp. 214-257.

Rosenthal, R., & Jacobson, L. *Pygmalion in the classroom.* New York: Holt, Rinehart & Winston, 1968.

Schull, W. J., & Neel, J. V. *The effects of inbreeding on Japanese children.* New York: Harper & Row, 1965.

Schwebel, M. *Who can be educated?* New York: Grove, 1968.

Scott, J. P., & Fuller, J. L. *Genetics and the social behavior of the dog.* Chicago: Univer. of Chicago Press, 1965.

Stodolsky, S. S., & Lesser, G. Learning patterns in the disadvantaged. *Harvard educ. Rev.*, 1967, **37**, 546-593.

Scrimshaw, N. S. Infant malnutrition and adult learning. *Saturday Review*, March 16, 1968. p. 64.

Shields, J. *Monozygotic twins brought up apart and brought up together.* London: Oxford Univer. Press, 1962.

Shuey, A. M. *The testing of Negro intelligence.* (2nd ed.) New York: Social Science Press, 1966.

Skeels, H. M. Adult status of children with contrasting early life experiences: A follow-up study. *Child Developm. Monogr.*, 1966, **31**, No. 3, Serial No. 105.

Skeels, H. M., & Dye, H. B. A study of the effects of differential stimulation on mentally retarded children. *Proc. Addr. Amer. Ass. Ment. Defic.*, 1939, **44**, 114-136.

Spuhler, J. N., & Lindzey, G. Racial differences in behavior. In J. Hirsch (Ed.), *Behavior-genetic analysis.* New York: McGraw-Hill, 1967. Pp. 366-414.

Stoch, M. B., & Smythe, P. M. Does undernutrition during infancy inhibit brain growth and subsequent intellectual development? *Arch. Dis. Childh.*, 1963, **38**, 546-552.

Stoddard, G. D. *The meaning of intelligence.* New York: Macmillan, 1943.

Stott, D. H. Interaction of heredity and environment in regard to 'measured intelligence.' *Brit. J. educ. Psychol.*, 1960, **30**, 95-102.

Stott, D. H. *Studies of troublesome children.* New York: Humanities Press, 1966.

Thompson, W. R. The inheritance and development of intelligence. *Res. Pub. Ass. Nerv. Ment. Dis.*, 1954, **33**, 209-331.

Thorndike, E. L. Measurement of twins. *J. Philos., Psychol., Sci. Meth.*, 1905, **2**, 547-553.

Tuddenham, R. D. Psychometricizing Piaget's méthode clinique. Paper read at Amer. Educ. Res. Ass., Chicago, February, 1968.

Tyler, L. E. *The psychology of human differences.* (3rd ed.) New York: Appleton-Century-Crofts, 1965.

U.S. Commission on Civil Rights. *Racial isolation in the public schools.* Vol. 1. Washington, D.C.: U.S. Government Printing Office, 1967.

U.S. News and World Report, Mental tests for 10 million Americans—what they show. October 17, 1966. Pp. 78-80.

Vandenberg, S. G. Contributions of twin research to psychology. *Psychol. Bull.,* 1966, 66, 327-352.

Vandenberg, S. G. Hereditary factors in psychological variables in man, with a special emphasis on cognition. In J. S. Spuhler (Ed.), *Genetic diversity and human behavior.* Chicago: Aldine, 1967. Pp. 99-133.

Vandenberg, S. G. The nature and nurture of intelligence. In D. C. Glass (Ed.), *Genetics.* New York: The Rockefeller University Press and Russell Sage Foundation, 1968.

Vernon, P. E. Symposium on the effects of coaching and practice in intelligence tests. *Brit. J. educ. Psychol.,* 1954, 24, 5-8.

Vernon, P. E. Environmental handicaps and intellectual development: Part II and Part III. *Brit. J. educ. Psychol.,* 1965, 35, 1-22.

Walters, C. E. Comparative development of Negro and white infants. *J. genet. Psychol.,* 1967, 110, 243-251.

Wheeler, L. R. A comparative study of the intelligence of East Tennessee mountain children. *J. educ. Psychol.,* 1942, 33, 321-334.

White, S. H. Evidence for a hierarchical arrangement of learning processes. In L. P. Lipsitt & C. C. Spiker (Eds.), *Advances in child development and behavior.* Vol. 2. New York: Academic Press, 1965. Pp. 187-220.

Willerman, L., & Churchill, J. A. Intelligence and birth weight in identical twins. *Child Developm.,* 1967, 38, 623-629.

Wilson, A. B. Educational consequences of segregation in a California community. In *Racial isolation in the public schools,* Appendices, Vol. 2 of a report by the U.S. Commission on Civil Rights. Washington, D.C.: U.S. Government Printing Office, 1967.

Wiseman, S. *Education and environment.* Manchester: Manchester Univer. Press, 1964.

Wiseman, S. Environmental and innate factors and educational attainment. In J. E. Meade & A. S. Parkes (Eds.), *Genetic and environmental factors in human ability.* New York: Plenum Press, 1966. Pp. 64-80.

Wright, S. Statistical methods in biology. *J. Amer. stat. Ass.,* 1931, 26, 155-163.

Young, M., & Gibson, J. B. Social mobility and fertility. In J. E. Meade & A. S. Parkes (Eds.), *Biological aspects of social problems.* Edinburgh: Oliver and Boyd, 1965.

Zazzo, R. *Les jumeaux, le couple et la personne.* 2 vols., Paris: Presses Universitaires de France, 1960.

Zigler, E. Familial mental retardation: a continuing dilemma. *Science,* 1967, 155, 292-298.

Zigler, E. The nature-nurture issue reconsidered: A discussion of Uzgiris' paper. Paper read at Conference on Sociocultural Aspects of Mental Retardation, Peabody College, Nashville, Tenn., June, 1968.

Discussion:
How Much Can We Boost IQ and Scholastic Achievement?*

The editors have solicited the following discussions in response to Professor Jensen's article, which appeared in the Winter, 1969, issue of the Review. *In his controversial article, Dr. Jensen develops a definition of the concept of intelligence and discusses the relative contribution of genetic and environmental influences in molding IQ. His conclusions run counter to many of the assumptions on which educational programs of the past few years have been based. For example, he argues that IQ is determined much more by genetic than by environmental influences. He argues that the most important environmental factors affecting intelligence occur prenatally and in the first year of life, and are associated mainly with the nourishment of mother and child. He analyzes the failure of many preschool and compensatory programs of the last five years to achieve significant gains in IQ and concludes that such programs are misdirected in their choice of goals and educational practices.*

In the following section six psychologists and a geneticist bring their disciplines and assumptions to bear on his arguments, examining the conceptual problem of estimating heritability, the relative effects of different environments on development, and Jensen's two-level conception of intelligence.†

* Arthur R. Jensen, "How Much Can We Boost IQ and Scholastic Achievement?," Harvard Educational Review, XXXIX (Winter, 1969), pp. 1-123.

† Since this Reprint Series was compiled, discussion has continued both in the pages of HER (Reprint Series No. 4 described on back cover) and elsewhere, for example, W. F. Bodmer and L. L. Cavalli-Sforza, "Intelligence and Race," *Scientific American*, Vol. 223, No. 4 (October, 1970), 19–29 and Jerome Hellmuth, ed., *Disadvantaged Child*, Vol. 3, *Compensatory Education: A National Debate* (New York: Brunner/Mazel, 1970).

Harvard Educational Review Vol. 39 No. 2 Spring 1969, 273

Inadequate Evidence and
Illogical Conclusions

JEROME S. KAGAN, *Harvard University*

Professor Kagan is critical of the logic of Dr. Jensen's article and presents evidence that any IQ data collected in the standardized manner may not reflect the actual potential of lower class children. In Kagan's opinion, Jensen's major fallacies are (1) his inappropriate generalization from within-family IQ differences to an argument that separate racial gene pools are necessarily different and (2) his conclusion that IQ differences are genetically determined, although he glosses over evidence of strong environmental influences on tested IQ—even between identical twins. Kagan cites new studies which suggest that part of the perceived intellectual inadequacy of lower class children may derive from a style of mother-child interaction that gives the lower class child less intense exposure to maternal intervention. Finally, Kagan argues, present compensatory education programs have been neither adequately developed nor evaluated. We cannot, therefore, use current evaluations of them to dismiss all possible compensatory programs.

Arthur Jensen's essay on IQ, scholastic achievement, and heredity contains a pair of partially correct empirical generalizations wedded to a logically incorrect conclusion. Professor Jensen notes first that scores on a standard intelligence test are more similar for people with similar genetic constitutions. The more closely related two people are, the more similar their IQ scores, suggesting that there is a genetic contribution to intelligence test performance. The second fact is that black children generally obtain lower IQ scores than whites. Unfortunately, Jensen combines the two facts to draw the logically faulted conclusion that there are genetic determinants behind the lower IQ scores of black children. The error in his logic can be illustrated easily, using stature as an example. There is no doubt that stature

Harvard Educational Review Vol. 39 No. 2 Spring 1969, 274–277

is inherited. Height is controlled by genetic factors. The more closely related two people are, the more similar their height. It is also true that Indian children living in the rural areas of most Central or South American countries are significantly shorter than the Indian children living in the urban areas of those countries. Jensen's logic would suggest that the shorter stature of the rural children is due to a different genetic constitution. However, the data indicate otherwise. The shorter heights of the rural children do not seem to be due to heredity but to disease and environmental malnutrition. The heights of children in many areas of the world, including the United States, have increased considerably during the past twenty years due to better nutrition and immunization against disease, not as a result of changes in genetic structure. Yet a person's height is still subject to genetic control. The essential error in Jensen's argument is the conclusion that if a trait is under genetic control, differences between two populations on that trait must be due to genetic factors. This is the heart of Jensen's position, and it is not persuasive.

Professor I. I. Gottesman, a leading behavioral geneticist, also questions the validity of Jensen's ideas. He notes that, ". . . even when gene pools are known to be matched, appreciable differences in mean IQ can be observed that could only have been associated with environmental differences." In a study of 38 pairs of identical twins reared in *different environments,* the average difference in IQ for these identical twins was 14 points, and at least one quarter of the identical pairs of twins reared in different environments had differences in IQ score *that were larger than 16 points.* This difference is larger than the average difference between black and white populations. Gottesman concludes, "The differences observed so far between whites and Negroes can hardly be accepted as sufficient evidence that with respect to intelligence the Negro American is genetically less endowed."

Let us consider some additional empirical evidence that casts doubt on the validity of Jensen's position. Longitudinal studies being conducted in our laboratory reveal that lower class white children perform less well than middle class children on tests related to those used in intelligence tests. These class differences with white populations occur as early as one to two years of age. Detailed observations of the mother-child interaction in the homes of these children indicate that the lower class children do not experience the quality of parent-child interaction that occurs in the middle class homes. Specifically, the lower class mothers spend less time in face to face mutual vocalization and smiling with their infants; they do not reward the child's maturational progress, and they do not enter into long periods of play with the child. Our theory of mental development suggests that specific absence of these experiences will retard mental growth and will lead to lower intelligence test scores. The most likely determinants of the black child's lower IQ score are his experiences during the first five years of life. These experiences lead the young black child to do poorly on IQ tests in part because he does not appreciate the nature of a problem.

A recent study of urban black children showed that the IQ distribution had two peaks. There was a large proportion of children with IQ scores around 60 and a much larger group whose distribution was normal and similar to that of white populations. The examiners felt that the very low IQ scores were a product of failure to understand the problem; failure to know what to do; failure to appreciate a test was being administered. This argument finds support in a recent study by Dr. Francis Palmer of the City University of New York. Dr. Palmer administered mental tests to middle and lower class black children from Harlem. However, each examiner was instructed not to begin any testing with any child until she felt that the child was completely relaxed, and understood what was required of him. Many children had five, six and even seven hours of rapport sessions with the examiner before any questions were administered. Few psychological studies have ever devoted this much care to establishing rapport with the child. Dr. Palmer found very few significant differences in mental ability between the lower and middle class populations. This is one of the first times such a finding has been reported and it seems due, in part, to the great care taken to insure that the child comprehended the nature of the test questions and felt at ease with the examiner.

We can quickly dismiss Jensen's suggestion that compensatory education is not likely to help black children. The value of Head Start or similar remedial programs has not yet been adequately assessed. It is not reasonable to assume that compensatory education has failed merely because eight weeks of a Head Start program organized on a crash basis failed to produce stable increases in IQ score. The flaws in this logic are overwhelming. It would be nonsense to assume that feeding animal protein to a seriously malnourished child for three days would lead to a permanent increase in his weight and height, if after 72 hours of steak and eggs he was sent back to his malnourished environment. It *may be* that compensatory education is of little value, but this idea has not been tested in any adequate way up to now.

Finally, it is important to realize that the genetic constitution of a population does not produce a specific level of mental ability; rather it sets a range of mental ability. Thus genetic factors are likely to be most predictive of proficiency in mental talents that are extremely difficult to attain, such as creative genius in mathematics or music, not relatively easy skills. Learning to read, write or add are easy skills, well within the competence of all children who do not have serious brain damage. Therefore, it is erroneous to suggest that genetic differences between human populations could be responsible for failure to master school related tasks. Ninety out of every 100 children, black, yellow or white, are capable of adequate mastery of the intellectual requirements of our schools. Let us concentrate on the conditions that will allow this latent competence to be actualized with maximal ease.

References

Gottesman, I. I. Biogenetics of race and class. In M. Deutsch, I. Katz, and A. R. Jensen (Eds.), *Social class, race, and psychological development.* New York: Holt, Rinehart & Winston, 1968.

Palmer, F., unpublished research reported at a colloquium at Harvard University, November, 1968.

Has Compensatory Education Failed?
Has It Been Attempted?

J. McV. HUNT, *University of Illinois*

While Professor Hunt finds much of interest in parts of Jensen's article, he objects strongly to some of its conclusions. Hunt fails to find satisfactory evidence that we may make the assertions about genetic differences determining the intelligence of Negroes and whites which Jensen has offered. He finds Jensen's claims about the high heritability of intelligence unsubstantiated; he finds Jensen's conclusion that observed group mean differences in IQ scores among Negro and white populations are genetically determined to be even less supportable. Hunt offers an alternative hypothesis: given the necessary relationship between the physical structure of the nervous system and the behavior of the system (as in IQ), we must provide rich post-natal experience in order to develop the inherent structures. He offers analogies from animal research which suggest that the physical development of the brain is directly influenced by its information-processing activities—these activities are particularly effective in neo-natal organisms.

Jensen's paper is a critical effort to correct the currently wide-spread "belief in the almost indefinite plasticity of intellect." He asserts that "the ostrich-like denial of biological factors in individual differences, and the slighting of the role of genetics in the study of intelligence can only hinder investigation and understanding of the conditions, processes, and limits through which the social environment influences human behavior" (p. 29). He finds my term "fixed intelligence" to be rather misleading for two real and separate reasons: (1) the genetic basis of individual differences in intelligence and (2) the stability or the constancy of the

Harvard Educational Review Vol. 39 No. 2 Spring 1969, 278–300

IQ throughout the individual's life. A major share of his paper is devoted to explaining the heritability of traits and to the theoretical and empirical basis for the proposition that about 80% of the individual variance in intelligence (defined in terms of the IQ and/or Spearman's *g*) has a genetic basis. This, at least by implication, explains why compensatory education "apparently has failed" (p. 2). He examines class differences and race differences in these same terms. But there is more to his paper. In the end, he offers, from the results of his own investigations, a basis for some hope through education if educational practice is modified.

Honest criticism is useful, both in science and in the process of social change which the behavioral, biological, and social sciences have now begun to influence. It is always useful unless it serves to hamper freedom of and support for investigation and for the development of appropriate technologies for coping with social problems. On the whole, Jensen's criticism comes in a constructive spirit. Moreover, it is informative. I am glad for the invitation to respond to his paper, for it has motivated more careful reading and consideration than I might otherwise have given it. In responding, I would like to synopsize his argument and respond point by point, but in the pages allowed me, I must respond selectively.

It is worth noting that Professor Jensen's argument is highly sophisticated in terms of both psychometrics and population genetics. His explanations in these domains are as briefly clear and as uncluttered with unnecessary jargon as any I have seen. He defines intelligence operationally in terms of what the IQ tests measure, of what accounts for the co-variation among test scores, (Spearman's *g*), and of the relations of these measures to scholastic ability (whence the tests come originally in the work of Binet and Simon), to occupational status, and to job success. What the IQ measures and what Spearman's *g* represents psychologically, he writes, "is probably best thought of as a capacity for abstract reasoning and problem-solving ability" (p. 19), and is also epitomized in cross-modality transfer. He recognizes clearly that intelligence is a phenotype, not a genotype:

. . . the IQ is not constant, but, like all other developmental characteristics, is quite variable early in life and becomes increasingly stable throughout childhood. By age 4 or 5, the IQ correlates about .70 with IQ at age 17, which means that approximately half [r^2] of the variance in adult intelligence can be predicted as early as age 4 or 5. (p. 18)

He does not note here that this increasing stability is based on a part-whole relationship wherein the IQs of successive ages constitute increasing proportions of IQ of the criterion age. He asks the traditional geneticists' question of how much variation (*i.e.*, individual difference) in measures of the intelligence phenotype of our population can be accounted for in terms of variation in genetic factors. He then presents the evidence for heritability (H) approximating .80 in European and North American Caucasian populations. Jensen explicitly accepts that the value of H holds only for the population sampled, and that under changed conditions the value of H could be expected to change.

Despite this psychometric and genetic sophistication in Professor Jensen's discourse, I find little evidence of an inclination to broaden the nomological net to include evidence from social psychology, from the physiological effects of early experience in animals, and from history to help interpret the psychometric and genetic findings. I find wanting an appreciation of how what Sumner (1906) called the "folkways" and Sherif (1936) has called the "social norms" can operate to produce radically different ecological niches for developing infants and children of differing social classes and races. I find wanting also an appreciation of individual lives as dynamic processes in which the preprogrammed information in the genetic code get cumulatively modified in both rate and direction by successive adaptations to the circumstances of the ecological niche. Thus, Professor Jensen's argument sums up to a sophisticated justification of what I have termed, and perhaps unfortunately, "fixed intelligence" and "predetermined development" (Hunt, 1961). Except for the educational significance he finds in the results of his own investigations, his argument allows only a eugenic approach to the problems of incompetence and poverty. With the exception of this loophole it is a counsel of despair, for our increasingly technological society cannot afford a century or two of selective breeding.

Points of Agreement

Even though my own theoretical predilections (or prejudices, perhaps) differ sharply from those of Professor Jensen, I have found many points in his paper with which I agree heartily. We agree, albeit for different reasons, that the concept of the "average child" is highly unfortunate in education. I find myself delighted with his thumbnail sketch of the central features of that traditional educational practice which has consequently evolved in Europe and America. It is the best I have ever seen. Unlike Jensen, however, I do not find imagining radically different forms so difficult even though I recognize that changing our educational folkways will be exceedingly difficult.

I agree that there is abundant evidence of genetic influences on behavior and

that one can increase or decrease by selective breeding the measures of any phenotypic trait which has been investigated, but I believe from evidence omitted in
Jensen's discourse that what Dobzhansky has termed the "range of reaction"
(Sinott, Dunn, & Dobzhansky, 1958, p. 22ff) is probably greater for intelligence
than it is for many other characteristics which depend less on what I suspect are
cumulative effects of successive adaptations.

I agree that it is essentially meaningless to speak of "culture free" and of "culture fair" tests, and yet I also agree that Cattell (1963) has made, on the basis of
differences within the intercorrelations, "a conceptually valid distinction between
two aspects of intelligence, *fluid* and *crystallized*" (p. 13).

I agree with Jensen that the technological advances in our culture make it
highly important to raise the intelligence, the educational attainments, and/or
the general competence of those people who now comprise the bottom quarter
of our population in measures of this cluster of characteristics. I agree that the
national welfare policies we established in the 1930s have probably operated in
disgenic fashion, and that it is highly important to establish welfare policies
which will encourage initiative and probably, in consequence, help foster positive
genotypic selection.

I could not agree more completely than I do with Professor Jensen's statement
that:

The variables of social class, race, and national origin are correlated so imperfectly with
any of the valid criteria on which [social] decisions [with respect to individuals] should
depend, or, for that matter, with any behavioral characteristic, that these background
factors are irrelevant as a basis for dealing with individuals—as students, as employees,
as neighbors. (p. 78)

Finally, for me the most interesting portion of Professor Jensen's paper is to be
found in the results of his own investigations. The absence of class differences in
what he calls "associative" learning, despite substantial differences in "cognitive" learning, is exceedingly interesting. Although I may well give a quite different interpretation of the basis for these findings than does Professor Jensen, I
agree equally strongly with the educational implication he draws from his findings. *One does not provide equality of educational opportunity by submitting
all children to the lock-step and by providing them with a single way in which to
develop their genotypic potential.* Variation in genotypes combines with variation in early experience to call for an increased individualization of education.
(Jensen's discussion is on pp. 6-8, 111-117.)

Points of Disagreement

Although I have found many points in Jensen's paper with which I can heartily agree, I have also found others with which I can just as heartily disagree. These are, first, several matters concerned with the measurement, the distribution, the development, and the nature of intelligence; second, the nature of his emphasis on biological versus psychological and social factors in behavioral development and the implications he draws for the relatively fixed nature of the existing norms for "intelligence." Third is Jensen's implicitly limited view of the learning process, coupled with his apparent lack of appreciation of the cumulative and dynamic implications of existing evidence of plasticity in the rate of behavioral development. Fourth are the implications which he draws for class and race differences from the measures of heritability of the IQ in European and American Caucasians. Finally, comes a disagreement about the wisdom of his opening sentence that "compensatory education has been tried and it apparently has failed" in the light of his avowed predilection for keeping all hypotheses open to investigation (and hopefully to technological development) as well as debate.

Matters Concerned with Intelligence

First, I find definitions of intelligence in terms of existing psychometric operations highly unsatisfying. Even though it was J. P. Guilford who introduced me to psychology and attracted me to the field largely with his discourse on aptitude testing and its implications for vocational guidance, I must confess that I have long distrusted the statistical operations of correlational analysis and averaging once they leave me without at least an intuitive connection with behavioral and biological observables. Thus, when Jensen remarks that Spearman's g-factor has "stood like a rock of Gibraltar," I find it hard to take seriously his avowance that "we should not reify g as an entity, of course, since it is only a hypothetical construct intended to explain covariation among tests" (p. 9). The g-factor explains on the average some 50% of the total variation in individual differences. Jensen notes further that "as the tests change, the nature of g will also change, and a test which is loaded, say, .50 on g when factor analyzed among one set of tests may have a loading of .2% or .8%, or some other value, when factor analyzed among other sets of tests" (p. 11). Apparently g is the most malleable and ameoboid rock extant. Jensen, however, makes a partial escape from his self-made operational *cul-de-sac* by arguing that intelligence is but one component of ability and competence. Thus, his own investigative finding that children of lower-class background can manage "associative" learning as well as children of middle-class background provides him with a ray of educational hope.

Professor Jensen devotes a substantial portion of his paper to an explication of the existing distribution of IQs in the population. He makes much of the basic normality of the distribution and the deviations from normality for pathological retardates and the "bulge" between 70 and 90 which he attributes to "the combined effects of severe environmental disadvantages and of emotional disturbances that depress test scores" (p. 27). Professor Jensen acknowledges that the traditional procedures provided by Binet and Simon for determining the mental age of any test-item forces the scores to assume a normal distribution, and he honestly admits that "the argument about the distribution of intelligence thus appears to be circular" (p. 21). He then argues that the only way out is to look for evidence that intelligence scales behave like an "interval scale." He finds the most compelling evidence from studies of the inheritance of intelligence. Am I emitting a mere flippancy if I respond that apparently, for Jensen, going twice around the circular argument removes its circularity? Actually, I find no serious fault with this discussion of the existing distribution of IQs in the population until Jensen begins to draw from it the implication that this existing distribution is fixed in human nature for all time, or until selective breeding alters it. My reasons for finding fault with this implication are derived from enlarging the nomological net to include evidence from outside the domains of psychometrics and population genetics as applied to intelligence, and I hope my argument will gradually become both clear and forceful.

On the matter of the stability of the IQ, Professor Jensen disavows any claim for constancy. On the other hand, he appears to view intellectual development as a matter of static, largely predetermined, growth. Thus he takes the findings of Bloom (1964) and emphasizes that half of the variance in the IQ at age seventeen can be predicted from IQs at ages of four and five years. If one considers the development of intelligence to be in substantial degree a function of the cumulative effects of informational and intentional interaction with physical and social circumstances, and if one takes into account the fact that the longitudinal predictive value of the IQ involves part-whole relationships, the emphasis can readily be reversed. Thus, just as embryologists have said that half of the epigenetic changes in a human life occur between conception and the end of the embryonic phase after only two months of gestation, it is more than a mere analogy to say that half of the epigenetic changes in mental development have typically taken place by about age four. This latter position puts the emphasis on the importance of early experience (including the intrauterine and nutritional) as both Bloom and I have been wont to do.

Perhaps I am wrong in inferring that Professor Jensen at least implicitly conceives a sharp distinction between tests of intelligence and tests of educational achievement, for he emphasizes that the former has substantially a higher heritability (80%) than the latter (approximately 60%). Because the main thrust of his paper is to emphasize the high heritability of intelligence, one can understand his omission of the papers by both Ferguson (1954, 1956, 1959) on the relation of learning to human ability and Humphreys (1962a, 1962b)[1] on the point that tests of intelligence and tests of academic achievement differ only in degree, in the sense that the former assess the results of incidental learning typically distant in time from that of the testing while the latter assess the results of learning in specific educational situations near in time to the testing. When one combines the evidence and arguments from these papers with a conception of intelligence as a cumulative, dynamic product of the ongoing informational and intentional interaction of infants and young children with their physical and social circumstances, one must call into question the notion of intellectual development as essentially a static function of growth, largely predetermined in rate.

The Dualism of Biological Versus Psychological (and Social) Factors

Professor Jensen quotes with high approval a paragraph by Edward Zigler to the effect that: "Not only do I insist that we take the biological integrity of the organism seriously, but it is also my considered opinion that our nation has more to fear from unbridled environmentalists than from those who point to such integrity as one factor in the determination of development.... It is the environmentalists who have placed on the defensive any thinker who, perhaps impressed by the revolution in biological thought stemming from discoveries involving DNA-RNA phenomena, has had the temerity to suggest that certain behaviors may be in part the product of read-out mechanisms residing within the programmed organism" (p. 29).

[1] Professor Hunt calls attention to research that was omitted in the pre-publication draft on which this discussion is based. The Humpreys data is included in the printed version of Jensen's article as a note on page 58. The reader's attention is directed to the opposite interpretations each author draws from the research. In effect, Hunt argues that the correlation of IQ and academic achievement indicates that IQ is dynamic and cumulative; Jensen holds to his conception of IQ as largely predetermined, and suspects that he has overestimated the malleability of academic achievement.

I believe that I have regularly taken "the biological integrity of the organism" seriously. Taking seriously the biological integrity of the organism is the major reason for my repeated concern with what I call "the problem of the match" between what has been built into the organism—through the program of maturation and through previous informational interaction with circumstances—and how newly encountered circumstances affect his motivation and continuing development (see Hunt, 1961, pp. 268-288; 1965; 1966, pp. 118-132). Also motivated by serious concern for the biological integrity of the organism is an extended effort to develop sequential ordinal scales of psychological development (Uzgiris & Hunt, 1969) and to look toward what one might term a "natural curriculum" for the fostering of early psychological development. In addition to these remarks, which may be regarded as defensive, it may be worth noting that the RNA (ribonucleic acid) phenomena are chiefly products of an organism's adaptation to circumstances.

Throughout his paper, and especially when he comes to the section on "how the environment works," the thrust of Professor Jensen's argument is to place psychological factors (and the social subset of these factors) in a kind of dualistic opposition to biological factors. Having implicitly constructed the dualism, he proceeds to denigrate the importance of the psychological set relative to the importance of biological set.

First, let me dispose of the dualism. Ample evidence has now accumulated to show that the consequences of informational interaction with circumstances, through the ears and the eyes (and especially the latter for the evidence extant), is quite as biological in nature as the effects of nutrition or of genetic constitution. Interaction through the eyes, especially early in life, has genuine neuroanatomical and neurochemical consequences.

Much of this evidence has its conceptual origin in the theorizing of Donald Hebb (1949). It was Hebb's hypothesis that the development of form-vision derives from sensory (S-S) integration that prompted Riesen and his colleagues to rear chimpanzees in the dark in order to determine the effect of light stimulation on the function and structure of the visual system. As is now widely known, a period of 16 or 18 months in total darkness produced drastic effects. On the functional side, there were a number of defects which proved essentially irreversible in those chimpanzees submitted to total darkness for 16 months or longer (see Riesen, 1958). On the side of anatomical structure, a defect was manifest during life as a pallor of the optic disc (Riesen, 1958). When these animals were sacrificed after some six years in full daylight, a histological examination brought out clear evidence of defects in the ganglion-cell layer of the retinae and in the optic nerve. These anatomical consequences within the visual system had themselves been irreversible (Chow, Riesen, & Newell, 1957). The histological examination also got evidence of a paucity of Mueller fibers within the retinal ganglia, and it should be noted that Mueller fibers are glia (Rasch, Swift, Riesen, & Chow, 1961).

Another line of investigation has stemmed from Hydén's (1961) biochemical hypothesis that memory and learning involve the metabolism of ribonucleic acid (RNA) in an interaction between neural and glial cells of the retina and brain. Hydén's hypothesis prompted Brattgård (1952) to rear rabbits in the dark. Histochemical analysis of the retinae of these dark-reared rabbits revealed a deficiency in RNA production of their retinal ganglion cells as compared with their light-reared litter-mates. Since then histological and histochemical effects of dark-rearing have been found not only in chimpanzees (Chow, et al., 1957) and rabbits, but also in kittens (Weiskrantz, 1958) and in rats (Liberman, 1962).

I have often expressed the wish that someone would extend this line of investigation centrally in the visual system to the lateral geniculate body of the thalamus and to the striate area of the occipital lobe. After regaling Robert Reichler of the National Institute of Mental Health with this evidence just outlined, I expressed again this wish to see an extension to the lateral geniculate body and to the striate area of the occipital lobe. Dr. Reichler responded excitedly that this had been done. In late October, he had attended an NIMH-supported conference on dyslexia where Dr. F. Valverde of Cajal's Institute in Madrid had presented a paper authored with Ruiz-Marcos which indeed reported such investigations with highly interesting findings. I am indebted to Dr. Reichler for letting me see a copy of the conference draft of the paper by Valverde and Ruiz-Marcos.

As yet I have had no opportunity to examine the evidence in detail, but their paper reviews an investigation by Wiesel and Hubel (1963), in which were described clearly evident defects in the cell areas of the lateral geniculate bodies on the thalami of kittens corresponding to the single eye deprived of vision for three months. Their paper also reviews evidence from investigations by Gyllesten (1959), by Coleman and Riesen (1968), by Ruiz-Marcos and Valverde (1968), by Valverde (1967, 1968), and by Valverde and Esteban (1968). All these investigations have shown clearly the effects of being reared in the dark, sometimes for only a very few days, on the fine structure of the striate area of the occipital lobe which is the center for visual reception. These effects show in the dendritic fields, and they show especially as a diminution in the number of spines on the dendrites of the large pyramidal cells in the striate area of the visual cortex (Valverde, 1967, 1968). Through electron-microscopy it was determined that the number of spines on these dendrites, in intervals at given distances from the wall of the cell body, is ordinarily very highly correlated with mouse age, but when mice are reared for various periods in the dark, this correlation is markedly diminished (Ruiz-Marcos & Valverde, 1968), and the diminution is especially marked for the days immediately after the eyes open. Clearly the psychological factor of dark-rearing produces neuro-anatomical and neurochemical effects not only in the eye but in the thalamus and in the visual area of the cortex. Thus, this psychological factor of visual function appears to be quite as biological in its consequences as are the consequences of nutrition and genotype.

Dark-rearing produces just the kind of anatomical effects one might envisage from Hebb's (1949) concepts of "cell assemblies" and "phase sequences." I see no reason to think that such processes should be less likely in human beings than in rodents. It takes little imagination, moreover, to extrapolate from these findings. I suspect that sensorimotor functioning, especially during the earliest phases of behavioral development in the first and second years, influences the development of such things as the spines on dendrites throughout the brain. The success of Hydén and Egyhazi (1962) in identifying with remarkable specificity the locus of the neuroanatomical and neurochemical effects of rats learning to climb a guy-rope suggests that each coordination, between vision-and-hand motion or between eye-function and ear-function, has its own neuro-electrical-chemical-anatomical equipment. I suspect that when such equipment has emerged as the consequence of a given bit of functional accommodation or learning, it can readily be employed in other functioning and thereby become the basis for the transfer of training. Moreover, as equipment has been developed in many domains, it can in all likelihood become one of the bases for the positive intercorrelation among tested abilities which Spearman called *g*.

In his section on "how the environment works" Professor Jensen contends that "below a certain threshold of environmental adequacy, deprivation can have a markedly depressing effect upon intelligence. But above this threshold, environmental variations cause relatively small differences in intelligence." He contends further: "The fact that the vast majority of the populations sampled in studies of the heritability of intelligence are above this threshold level of environmental adequacy accounts for the high values of the heritability estimates and the relatively small proportion of IQ variance attributable to environmental influences" (p. 60). The evidence of increase in the development of brain structures following enrichments of early experience are hardly consonant with this position. Altman and Das (1964), for instance, have reported a higher rate of multiplication of glial cells in the cerebral cortices of rats reared in "enriched environments" and in rats reared in the "impoverished environments" of laboratory cages. In another extended program of such investigation which has been underway for more than a decade at the University of California, Bennett, Diamond, Krech, and Rosenzweig (1964) and Krech, Rosenzweig, and Bennett (1966) have done a long series of studies which indicate that rats reared in relatively complex environments have shown cortical tissue greater in weight and thickness than that of litter-mates reared in the simpler environments of laboratory cages. Here "complexity" has been defined in terms of the variety of objects available for the rats to perceive and to manipulate and the variety of different kinds of space to be explored. These rats reared in complex environments have also shown histochemical effects in the form of higher total acetylcholinesterase activity of the cortex than the cage-reared rats. Associated with these neuroanatomical and neurochemical effects of the life history, moreover, is a higher level of maze-problem-solving ability in the rats reared under complex circumstances than in those reared in laboratory cages.

The definition "of a certain threshold of environmental adequacy" is unclear, but it can be said that cage-rearing is the standard ecological niche of laboratory rats and that it involves no serious absence of light and sound. Contrary to Jensen's position that it is only below "a certain threshold of environmental adequacy" that there can be a markedly depressing effect on intelligence, I am inclined to suspect that the basic central equipment for the inter-modal transfer which Jensen conceives to be a prime example of Spearman's g can be greatly modified by the informational interaction of the human infant and young child with his physical and social circumstances. I say that I suspect this is the state of affairs. This statement has not been proven, but the thrust of the existing evidence points strongly in the direction which I have indicated.

Learning and the Cumulative Implication of Plasticity in Early Development

The traditional view of heredity and environment held them to be essentially separate processes in development, and maturation was conceived to be the developmental representative of heredity, with learning the developmental representative of environment. We have just seen that the young organism's adaptations to the environment influence maturation, but we have not clarified the nature of learning.

Learning is typically conceived in terms of the ways it has been investigated in the laboratory. Investigations of learning still bear the marks of the pioneers: Ebbinghaus for rote learning, Bryan and Harter for skill learning, Pavlov for classical conditioning, and, for the fourth general category, C. Lloyd Morgan and E. L. Thorndike for trial-and-error with reinforcement, Clark L. Hull for instrumental learning motivated by drive and reinforced by drive-reduction, and B. F. Skinner for operant conditioning. If one examines the developmental observations of Piaget (1936, 1937), wherein accommodation and assimilation become the terms for learning, one finds several kinds of effects of encounters with circumstances which have failed to get investigated in psychological laboratories. If one examines the almost forgotten work on attention, the work of the ethologists, and the work of social psychologists on attitude change and communication, one finds other kinds of modification of function, and presumably of neuroanatomy and neurochemistry, through encounters with informational circumstances which do not get into the chapters on learning. I believe I have identified eight kinds of learning seldom studied for themselves which appear to be operative in psychological development (Hunt, 1966). The number is unimportant; the point is that Professor Jensen's distinction between associative learning and cognitive learning is but a conceptual drop in the bucket. His finding that the class-differences evident for cognitive learning are not evident for associative learning is exceedingly interesting, however.

What appears to be wrong with Professor Jensen's implicit conception of learning is that it consists only (or basically) of those minor changes of function which can be effected within short intervals of time in the laboratory. Thus, he speaks of learning ability as a kind of static trait which accounts for the number of trials required for the assimilation or mastery of relatively miniscule accommodations.

Except for the case where he calls for studies of the transfer of learning before age five to the cognitive functions after age six (in which I join him), I miss in his discourse any strong appreciation of what must be the cumulative dynamic effects of adaptations at one phase of development on the adaptations of later phases. Thus, he can write of the influence of the genotype "reading through the environmental overlay."

Although Professor Jensen acknowledges that such "extreme sensory and motor restrictions in environments such as those described by Skeels and Dye (1939) and Davis (1947), in which the subjects had little sensory stimulation of any kind and little contact with adults" (p. 60) resulted in large deficiencies in IQ, he tends to minimize their importance. He notes in favor of his view that the orphanage children of Skeels and Dye gained in IQ from an average of 64 at 19 months of age to 96 at age six as a result of being given "social stimulation and placement in good homes at between two and three years of age" (p. 60). He notes that when these children were followed up as adults, they were found to be average citizens in their communities, and their own children had an average IQ of 105 and were doing satisfactorily in school. Similarly, Davis (1947) reported the more extreme case of Isabel, who had an IQ of 30 at age six, but who, when put into an intensive educational program at age six, developed a normal IQ by age eight. From these examples, he contends that even extreme environmental deprivation need not permanently result in below-average intelligence.

Professor Jensen neglects to report the results of the follow-up study of the adult status of the Skeels-Dye children left in the orphanage (Skeels, 1966). Those who were removed from the orphanage before they were 30 months old and placed on a women's ward at a state institution for the mentally retarded, and then later adopted, were all self-supporting and none became a ward of any institution. Their median educational attainment was 12th grade. Four had one or more years of college work, one received a bachelor's degree and went on to graduate school. On the other hand, of the 12 children who remained in the orphanage, one died in adolescence following continued residence in a state institution for the mentally retarded, and four remained on the wards of such institutions. With one exception, those employed were marginally employed, and only two had married. It is true that the effects of early experience can be reversed; the point to be made here, however, is that the longer any species of organism remains under any given kind of circumstances, the harder it is to change the direction of the effects of adaptation to those circumstances.

Even in infants reared in middle-class homes evidence exists of a remarkable degree of plasticity in early behavioral development. In my own laboratory, for instance, Greenberg, Uzgiris, and Hunt (1968) have shown that putting an attractive pattern over the cribs of such infants beginning when they are five weeks old, reduces the age at which the blink-response becomes regular for a target-drop of 11.5 inches from a mean of 10.4 weeks, in infants whose mothers agreed to put nothing over the cribs of their infants for 13 weeks, to a mean of 7 weeks. In the familiar terms of the IQ ratio this represents an increase of 48 points for the blink-response. The differences between the groups in mean age for drops of 7 inches and for drops of 3 inches becomes progressively less. Thus, the findings are quite consonant with those studies of the twenties and thirties which found the effects of practice on given skills to be evanescent. On the other hand, if one provides circumstances which permit the hastened looking schema, indicated by the blink-response, to be incorporated into a more complex sensorimotor organization, its early availability should be reflected in increased advancement. This is precisely the sort of thing one finds in the work of White and Held (1966). In their work, the capacity for visual accommodation permits looking to become incorporated into eye-hand coordination. In a normative study of successive forms of eye-hand coordination, top-level reaching failed to appear until the median age of the group was 145 days. The second enrichment program reduced the median age for top-level reaching from this 145 days to 87 days—an advance of 66 points in the familiar terms of the IQ ratio for this final level of eye-hand coordination. Hypothetically, at least, one should be able to extrapolate on this principle, but as yet experimental evidence is unavailable to confirm the hypothesis.

Cumulative and dynamic implications of this existing evidence of plasticity in the rate of behavioral development raises the question of what Dobzhansky has termed the "norm of reaction" (see Sinnott, Dunn, & Dobzhansky, 1958, p. 22ff) for the case of human intelligence. Although no one can now say how large the cumulative modifications in measurements of human intelligence might possibly be, Wayne Dennis (1966) has published a study which is highly relevant. The study examines the mean IQs from the Draw-a-Man Test for groups of typical children aged six and seven years from some 50 cultures over the world. Florence Goodenough (1926) devised this test to be culture free. Its freedom from cultural influences was called into question, however, when typical Hopi Indian children of six and seven turned up with a mean of 124 on the test (Dennis, 1942). This mean of 124 equaled the mean IQ for samples of upper-middle-class suburban American children and for samples of children from Japanese fishing villages. The lower end of this distribution of mean IQs finds nomadic Bedouin Arab children of Syria with a mean IQ of 52. Here, then, we find direct evidence of a norm of reaction of about 70 points in Draw-a-Man IQ. The most obvious correlate of this variation in mean IQ is amount of contact with the pictorial art. Among Moslem Arab children, whose religion prohibits representative art as graven images, the range in mean Draw-a-Man IQ is from 52 to 94, and the most obvious correlate of this norm of reaction is contact with groups of the Western culture. This is the most direct evidence concerning the norm of reaction for human intelligence of which I know. While the factor structure of the Draw-a-Man Test is probably considerably less complex than is that of either the Stanford-Binet or the Weschler Children's Scale, within our own culture Draw-a-Man scores correlate about as well with those from these other more complex scales as scores on them to with each other.

In connection with this discussion of the norm of reaction, which Professor Jensen mentions but to which he gives little attention, it is interesting to note what he omits from a paragraph quoted from the geneticist, Dobzhansky (1968b, p. 554 quoted in Jensen, p. 30). The omitted portion reads: "Although the genetically-guaranteed educability of our species makes most individuals trainable for most occupations, it is highly probable that individuals have more genetic adaptability to some occupations than to others. Although almost everybody could become, if properly brought up and properly trained, a fairly competent farmer, or a craftsman of some sort, or a soldier, sailor, tradesman, teacher, or priest, certain ones would be more easily trainable to be soldiers and others to be teachers, for instance. It is even more probable that only a relatively few individuals would have the genetic wherewithal for certain highly specialized professions, such as musician, or singer, or poet, or high achievement in sports or wisdom or leadership."

Finally, I am among those few who are inclined to believe that mankind has not yet developed and deployed a form of early childhood education (from birth to age five) which permits him to achieve his full genotypic potential. Those studies which so sharply disconfirmed what R. B. Cattell (1937) once characterized as a "galloping plunge toward intellectual bankruptcy," (see Hunt, 1961, p. 337ff) can probably be repeated again after 20 to 25 years if our society supports the necessary research and development of educational technology to enable us to do early childhood education properly. In connection with this possibility of a general increase in intelligence, we should consider also what has happened to the stature of human beings. It appears to have increased by nearly a foot without benefit of selective breeding or natural selection. While visiting Festival Park in Jamestown, Virginia recently, we examined the reproductions of the ships which brought the settlers from England. They were astoundingly small. The guide reported that the average height of those immigrants was less than 5 feet, and that the still famous Captain John Smith was considered to be unusually tall at 5 feet 2 inches. The guide's "instruction book" puts the authority for these statements in the Sween Library at William and Mary. I have been unable to check the evidence, but scrutiny of the armor on display in various museums in England implies that the stature of the aristocrats who wore it must typically have been about the reported size of those immigrants to Jamestown. Also, the guide for the U. S. *Constitution* includes in his spiel the statement that the headroom between decks needed to be no more than 5 feet and 6 inches because the average stature of sailors in the War of 1812 was but 5 feet and 2 inches. This increase in height can occur within a single generation. Among the families of German Russians whom I knew while growing up in Nebraska it was typical to find the average height of the children several inches above mid-parent height, and I can cite instances in which the increase was approximately a foot where all the children were sons. Inasmuch as Professor Jensen resorts repeatedly to the analogy between intelligence and stature, such evidence of an increase in the average height for human beings, the reasons for which are still a matter largely of conjecture, should have some force in increasing the credibility for the genetic potentiality for a general increase in intelligence.

Implications from Existing Measures of Heritability

Professor Jensen recognizes explicitly that measures of heritability may change as the nature of the population changes. Nevertheless, from these existing measures of heritability in European and American Caucasians, he draws implications for both class and race differences which, in view of the considerations already presented, I simply cannot accept at face value.

From the physiological evidence, from the fact that one can readily hasten the development of sensorimotor organizations in children of the middle class, and from the fact that technological advances have quite regularly increased the mean IQ of populations, I see no reason to believe that the current distribution of intelligence is fixed by the biological nature of man, despite the fact that heritability studies indicate that approximately 80% of the individual variance in the IQ can be attributed to variations in genotypes. Moreover, in view of the sharp contrast between the child-rearing practices of the middle class with those of the people of poverty, I see no reason to believe that the class differences now evident are inevitable. Finally, inasmuch as black people have had more than a century in slavery and then, since the war between the States, another century in both poverty and the bondage of "folkways," I see no reason to consider existing race differences as inevitable.

The contrast between the child-rearing of the middle class and that of the poor needs to be better understood. A study by Maxine Schoggen at the Demonstration and Research Center for Early Education at the George Peabody College for Teachers in Nashville is bringing out this contrast more forcefully than any other of which I know. The studies concern samples of eight families of professional status, eight of rural poverty, and eight of urban poverty. The families of rural poverty are white; those of urban poverty are black. In each family there is a 3-year-old who is the target-child. Observer-recorders become so well acquainted with these families that they become like furniture. They record for equal periods of time in functionally equivalent situations like meal time, bed time, and the time when the older children return from school. The observers record the instances of social interaction initiated by the older members of the family toward the target-child, and their reactions to the interaction initiated by the child. These are termed "environmental force units." From the evidence available, the older members of professional families initiated somewhat more than twice as many "environmental force units" per unit of time toward the 3-year-old in their family as did the older members in the families of either urban or rural poverty. I have asked Dr. Schoggen about how much difference there might be in the frequency of units in which the older members of the family would call upon a child to note the shape, the size, the color, or even the placement of objects and persons. She has indicated that this is quite common in professional families, but that it seldom occurs in the families of poverty except in connection with errands. Then the child usually gets castigated for his stupidity. On the verbal side, professional parents often call upon their three-year-olds to formulate such matters in language of their own, but families of either rural or urban poverty almost never do. One should recall in this connection that "warm democratic" rearing was associated with an average gain of 8 IQ points, over a three year period between the ages of approximately three or four to six or seven, in the study of Baldwin, Kalhorn, and Breese (1945), while the mean IQ dropped a point or two in the children of parents employing what these authors characterized as "passive-neglectful" and "actively-hostile" child-rearing (Baldwin, 1955, p. 523). This contrast between the rearing practices in families of professional status with those in families of either rural or urban poverty appears to be sharper than that between the families utilizing the various kinds of child-rearing identified by Baldwin, *et al*. Few if any of the studies of heritability have included the truly poor, so they have missed this portion of the variation in the circumstances of rearing.

At least a substantial portion of parents of poverty can be taught, however, to be effective teachers of their young when they are given models to imitate, when the actions of the models are explained, and when home visitors are provided to bring the new ways of child-rearing into the home (Gordon and associates, 1969; Karnes, 1969; Klaus & Gray, 1968; Miller, 1968). Moreover, when parents are involved in the education of their young children, they communicate new-found practices to their neighbors and the parents themselves take a new lease on life. In the Karnes project, the mothers agreed that if they were to give each child the attention needed, they dare not have a new baby each year, and so they all enrolled at the local Planned Parenthood Clinic. Miller (1968) reports that in the extension of the Early Training Project a majority of the mothers have upgraded their skills, and the families in the projects have formed clubs—one in which husbands and wives bowl regularly.

It will be no easy matter to spread this kind of training to all the families of poverty throughout this country, but a start has been made in the Parent and Child Centers which the Office of Economic Opportunity has established on a pilot basis.

The enrolling in the Planned Parenthood clinic suggests that this kind of enterprise in early childhood education instigates help to prevent some of the disgenic processes with which Professor Jensen and I are both concerned.

I applaud Professor Jensen's proposal to develop a curriculum based upon his finding that children of lower-class background are equal in "associative" learning to children of middle-class background. In doing so, he may ultimately help to raise the general level of competence, and even the intelligence defined as Spearman's g, in the next generation of those who receive the benefit of his efforts to develop new educational technology. Moreover, since the effects of early experience can be reversed, at least in part, if and when Professor Jensen builds educationally upon the capacity of children from lower-class background for "associative" learning, he will probably increase measures of their g-factor gradually. His program will also probably increase measures of Cattell's "crystallized" intelligence in his pupils. To a lesser degree his program may also increase measures of their "fluid" intelligence. Moreover, Professor Jensen's program could well contribute to an increase in the intelligence of the next generation.

If one views societal evolution as a process, the mean of the IQ on the basis of existing standardizations and the existing measures of heritability can well be seen as the pre-measures to be compared with post-measures (based in the case of the IQ, of course, on today's standardizations) 10 or 20 years hence.

The Opening Sentence

At one point in his paper, Professor Jensen makes an ardent plea for keeping all hypotheses open for debate and investigation. With this plea, I heartily agree.

Unfortunately, since social change is a process, one cannot settle the issue between my reading of the broad range of evidence and his reading of contemporary evidence from existing distributions of IQs and contemporary measures of heritability, until these changes in the ecological niche of infants and young children, to be accomplished by the research, the development, and the deployment of early childhood education, have been available for at least a decade or two. Saying outright that "compensatory education has been tried and it apparently has failed" is but a half-truth. Moreover, it is but a half-truth which can help to boost the forces of reaction which could halt support for research on how to foster psychological development, for the development of technology of early childhood education, and for the deployment of that technology across the USA. Insofar as it succeeds in boosting these forces of reaction, it will leave the issue open only for debate. Once the support for investigation, development, and deployment has been removed, the differences between our readings of the evidence will no longer be open for "investigation.

Perhaps I should explain why Professor Jensen's sentence is but a half-truth. In this sentence, "compensatory education" implies Head Start, for it is Head Start which has been tried—at least a little. Project Head Start did deploy a form of early childhood education for which many had hopes of compensatory effects. It was hoped that giving children of the poor a summer or two or a year of nursery school, beginning at age four, would overcome the handicaps of their earlier rearing. I hoped it would, but I feared from the beginning that such broad deployment of a technology untested for the purpose might lead to an "oversell" which, with failure of the hopes, would produce an "overkill" in which would be lost, for who knows how long, the opportunity to bring into the process of social change, in the form of early childhood education, the implications of the various lines of evidence indicating the importance of early experience for intellectual development. The 1967 report of the U. S. Commission of Civil Rights is correct in stating that Head Start has not appreciably raised the educational achievement of the children who participated. There is, however, a reason which absolves compensatory education as such.

Maria Montessori in Italy and Margaret McMillan of England established nursery schools to aid the children of the poor. These were brought to America along with the intelligence tests and just as the emphasis on learning by doing was becoming established. Nursery schools did not survive in America as aids for children of the poor. Rather, they got adapted for the children of the well-to-do who could pay for them. Moreover, when the psychoanalytic movement coalesced with Froebel's kindergarten movement and with the Child Study movement of G. Stanley Hall, the goal became one of freeing young children, for at least part of each day, from their mothers' strict disciplinary controls. Free play became the mode. Since such nursery schools constituted the only early education model available when Project Head Start began, traditional nursery school practice was the kind of early education deployed for the most part.

But Head Start is not synonymous with compensatory education. Professor Jensen knows this for he reviews a number of the investigations of compensation in one of the later sections of his paper. Compensatory education has not failed. Investigations of compensatory education have now shown that traditional play school has little to offer the children of the poor, but programs which made an effort to inculcate cognitive skills, language skills, and number skills, whether they be taught directly or incorporated into games, show fair success. A substantial portion of this success endures. If the parents are drawn into the process, the little evidence available suggests that the effect on the children, and on the parents as well, increases in both degree and duration. All this in seven years sounds to me like substantial success. Yet, we still have a long way to go before we shall have learned what an appropriate curriculum for infants from birth to five might be. Thus, Jensen's opening statement is a half-truth, and a dangerous half-truth, placed out of context for dramatic effect.

Insofar as the behavioral and educational sciences get involved in manning the tiller of social change, the practitioners of these sciences must learn to think in terms of processes and they must learn to think of political and social consequences of how and what they write and say. It does no good to plead for keeping all hypotheses open for debate and investigation if the form of the debate removes support for the relevant investigation and for the development and deployment required for a meaningful test of the hypotheses. I find it hard to forgive Professor Jensen for that half-truth placed out of context for dramatic effect at the beginning of his paper.

How much *can* we boost IQ and scholastic achievement by deliberately altering the ecological niche of infants and young children, from birth to age five, through early childhood education? Who knows? As I read the evidence, the odds are strong that we can boost both IQ and scholastic achievement substantially, but we cannot know how much for at least two decades. Moreover, we shall never find out if we destroy support for the investigation of how to foster early psychological development, for the development of educational technology, and for the deployment of that technology.

References

Altman, J. & Das, G. D. Autoradiographic examination of the effects of enriched environment on the rate of glial multiplication in the adult rat brain. *Nature*, 1964, **204**, 1161-1165.

Baldwin, A. L. *Behavior and development in childhood.* New York: Dryden Press, 1955.

Baldwin, A. L., Kalhorn, J. & Breese, F. H. Patterns of parent behavior. *Psychological Monographs*, 1945, **58**, No. 3, 1-75.

Bennett, E. L., Diamond, M. C., Krech, D., & Rosenzweig, M. R. Chemical and anatomical plasticity of the brain. *Science*, **146**, No. 3644 (1964), 610-619.

Bereiter, C., & Engelmann, S. *Teaching disadvantaged children in the preschool.* New York: Prentice Hall, 1966.

Bloom, B. S. *Stability and change in human characteristics.* New York: John Wiley & Sons, Inc., 1964.

Brattgård, S. O. The importance of adequate stimulation for the chemical composition of retinal ganglion cells during early post-natal development. *Acta Radiological* (Stockholm), 1952, Supplement 96, 1-80.

Cattell, R. B. *The fight for our national intelligence.* London: King, 1937.

Cattell, R. B. Theory of fluid and crystallized intelligence: a critical experiment. *Journal of Education Psychology*, 1963, **54**, 1-22.

Chow, K. L., Riesen, A. H., & Newell, F. W. Degeneration of retinal ganglion cells in infant chimpanzees reared in darkness. *Journal of Comparative Neurology*, 1957, **107**, 27-42.

Coleman, P. D., & Riesen, A. H. Environmental effects on cortical dendritic fields. I. Rearing in the dark. *Journal of Anatomy* (London), 1968, **102**, 363-374.

Davis, K. Final note on a case of extreme isolation. *American Journal of Sociology*, 1947, **57**, 432-457.

Dennis, W. The performance of Hopi Indian children on the Goodenough Draw-a-Man Test. *Journal of Comparative Psychology*, 1942, **34**, 341-348.

Dobzhansky, T. On genetics, sociology, and politics. *Perspectives in Biology and Medicine*, 1968, **11**, 544-554.

Ferguson, G. A. Learning and human ability: A theoretical approach. *Factor analysis and related techniques in the study of learning.* Edited by P. H. DuBois, W. H. Manning, & C. J. Spies. A report of a conference held at Washington University in St. Louis, Missouri, February, 1959. Technical Report No. 7, Office of Naval Research Contract No. Nonr 816 (02).

Goodenough, F. L. *The measurement of intelligence by drawings.* Yonkers-on-Hudson, New York: World Book Co., 1926.

Gordon, I. J. (ed.) *Reaching the child through parent education: The Florida approach.* Gainesville: Institute for Development of Human Resources, College of Education, University of Florida, 1969.

Greenberg, D., Uzgiris, I., & Hunt, J. McV. Hastening the development of the blink-response with looking. *Journal of Genetic Psychology*, 1968, **113**, 167-176.

Gyllensten, L. Postnatal development of the visual cortex in darkness (mice). *Acta Morphologica* (Neerlando-Scandinavica), 1959, **2**, 331-345.

Hebb, D. O. *The organization of behavior.* New York: John Wiley & Sons, Inc., 1949.

Humphreys, L. G. The nature and organization of human abilities. *The 19th Yearbook of the National Council on Measurement in Education.* Edited by M. Katz. Ames, Iowa, 1962a.

Humphreys, L. G. The organization of human abilities. *American Psychologist,* 1962b, **17**, 475-483.

Hunt, J. McV. *Intelligence and experience.* New York: Ronald Press, 1961.

Hunt, J. McV. Intrinsic motivation and its role in psychological development. *Nebraska Symposium on Motivation,* **13**. Edited by D. Levine. Lincoln: University of Nebraska Press, 1965.

Hunt, J. McV. Toward a theory of guided learning in development. In *Giving emphasis to guided learning.* Edited by R. H. Ojemann & K. Pritchett. Cleveland, O.: Educational Research Council, 1966.

Hunt, J. McV. Poverty versus equality of opportunity. In *The challenge of incompetence and poverty.* Urbana: University of Illinois Press (in press).

Hydén, H. Biochemical aspects of brain activity. *Man and civilization: Control of the mind.* Edited by S. M. Farber & R. H. L. Wilson. New York: McGraw-Hill, 1961.

Hydén, H., & Egyhazi, E. Nuclear RNA changes of nerve cells during a learning experiment in rats. *Proceedings of the National Academy of Science,* 1962, **48**, 1366-1373.

Karnes, M. B. *A new role for teachers: Involving the entire family in the education of preschool disadvantaged children.* Urbana: University of Illinois, College of Education, 1969.

Klaus, R. A., & Gray S. The early training project for disadvantaged children: A report after five years. *Monographs of the Society for Research in Child Development,* 1968, **33**, No. 4.

Krech, D., Rosenzweig, M. R., & Bennett, E. L. Environmental impoverishment, social isolation, and changes in brain chemistry and anatomy. *Physiology and Behavior.* London: Pergamon Press, 1966.

Liberman, R. Retinal cholinesterase and glycolysis in rats raised in darkness. *Science,* 1962, **135**, 372-373.

Miller, J. O. *Diffusion of intervention effects.* Urbana: ERIC Clearinghouse for Early Childhood Education, University of Illinois, 1968.

Rasch, E., Swift, H., Riesen, A. H., & Chow, K. L. Altered structure and composition of retinal cells in dark-reared mammals. *Experimental Cellular Research,* 1961, **25**, 348-363.

Riesen, A. H. Plasticity of behavior: Psychological aspects. In H. F. Harlow & C. N. Woolsey (Eds.), *Biological and biochemical bases of behavior.* Madison: University of Wisconsin Press, 1958, 425-450.

Ruiz-Marcos, A., & Valverde, F. Mathematical model of the distribution of dendritic spines in the visual cortex of normal and dark-raised mice. *Journal of Comparative Neurology* (in press).

Sinnott, E. W., Dunn, L. C., & Dobzhansky, T. *Principles of genetics.* New York: Mc-Graw-Hill, 1958.

Sherif, M. *The psychology of social norms.* New York: Harper, 1936.

Skeels, H. M. Adult status of children with contrasting early life experiences. *Monographs of the Society for Research in Child Development,* 1966, **31** (No. 3), 1-66.

Skeels, H .M., & Dye, H. B. A study of the effects of differential stimulation of mentally retarded children. *Proceedings of the American Association of Mental Deficiency,* 1939, **44,** 114-136.

Sumner, W. G. *Folkways.* Boston: Ginn & Co., 1906 (1940).

Uzgiris, I. C., & Hunt, J. McV. *Toward ordinal scales of psychological development in infancy* (in press).

Valverde, F. Apical dendritic spines of the visual cortex and light deprivation in the mouse. *Experimental Brain Research,* 1967, **3,** 337-352.

Valverde, F. Structural changes in the area striata of the mouse after enucleation. *Experimental Brain Research,* 1968, **5,** 274-292.

Valverde, F., & Esteban, M. E. Peristriate cortex of mouse: Location and the effects of enucleation on the number of dendritic spines. *Brain Research,* 1968, **9,** 145-148.

Weiskrantz, L. Sensory deprivation and the cat's optic nervous system. *Nature,* 1958, **181,** 1047-1050.

White, B. L., & Held, R. Plasticity of sensorimotor development in the human infant. In J. F. Rosenblith & W. Allinsmith (Eds.), *The causes of behavior: Readings in child development and educational psychology* (2nd edition). Boston: Allyn & Bacon, 1966.

Wiesel, T. N., & Hubel, D. H. Effects of visual deprivation on morphology and physiology of cells in the cat's lateral geniculate body. *Journal of Neurophysiology,* 1963, **26,** 978-993.

Genetic Theories and Influences:
Comments on the Value of Diversity

JAMES F. CROW, *University of Wisconsin*

Professor Crow agrees "for the most part with Jensen's analysis." He does suggest qualifications when drawing inferences from existing studies in biometrical genetics. First, he notes his reservations about the reality of the mathematical assumptions implicit in analysis of variance models. Second, he draws attention to the limited sample size available in studies of twins and siblings reared apart and asks how representative such groups are. Third, he notes that predictive models have inherent limits when new, qualitatively different, treatments are introduced into the environment.

Biometrical genetics has become quite a sophisticated subject with a substantial body of mathematical theory. One reason for this development is that the simple, mechanistic nature of Mendelian inheritance is very inviting to probabilistic model-builders. A second reason is that, because complex traits depend on more genes than can be individually identified, some sort of statistical treatment is necessary.

The foundations for the theory of biometrical genetics were laid by R. A. Fisher and Sewall Wright, using methods depending mainly on correlation and variance analysis. The procedures are widely used in animal breeding, thanks especially to J. L. Lush, who has been conspicuously successful in adapting these procedures to use in practical breeding problems. The general theory and methodology in this field are described with insight and lucidity by Falconer (1960).[1] Jensen's article, together with many others that he has written recently on this subject (see his bibliography), constitutes a thorough review and synthesis of

[1] This and subsequent references are to articles and books cited in Jensen's article.

the various attempts to apply these methods to human intelligence and scholastic achievement. Jensen has become a leader in this field, and I as a population geneticist, admire his understanding of the methods and his diligence and objectivity in bringing together evidence from diverse sources. He presents the evidence fairly, relying on empirical data in preference to introspection or traditional wisdom, and is very careful to distinguish between observation and speculation.

I shall confine my comments mainly to the genetic aspects of the article. I agree for the most part with Jensen's analysis. Any differences could probably best be described by saying that, in general, I have somewhat less confidence than he in the quantitative validity of the methods—more reservations about the reality of the necessary assumptions. I don't mean by this that I would reach opposite conclusions; I am simply more agnostic. This is especially true as regards intergroup comparisons and, in particular, the importance of genetic factors in racial differences.

The Concept of Heritability

Much of Jensen's article is concerned with the heritability of intelligence (I am trying to use the term intelligence in the same technical sense as he does). The word heritability has been used for some time by psychologists studying twins, but the measures—such as Holzinger's H-index and various modifications thereof—have not usually corresponded to the geneticist's definition. Jensen has done a great deal to clarify this point.

Heritability, in the geneticist's terms, can be described in three equivalent ways, depending on whether the viewpoint is that of analysis of variance, regression, or correlation analysis: (1) the ratio of the genetic variance to the total variance, (2) the regression of genotype on phenotype, and (3) the square of the correlation of genotype with phenotype. Jensen uses mainly the first.

As he says, the total or phenotypic variance (V_P) can be analyzed into genotypic (V_H) and environmental (V_E) fractions:

$$V_P = V_H + V_E + \text{interactions and error.}$$

The genotypic variance (V_H) can be subdivided further into the additive or genic variance (V_G), dominance (V_D), and inter-locus interaction (epistasis) (V_I). (See Jensen, p. 37.) Thus,

$$V_H = V_G + V_D + V_I.$$

V_G is defined as the best linear representation of the phenotypic values (best in the least squares sense), and V_D and V_I are treated as deviations from it. This procedure for subdividing V_H has two important advantages: The first is that V_G, V_D, and V_I defined this way are independent and we do not have to worry about covariances among them.

The second advantage is that V_G provides a means for predicting future generations when there is selection. The reason for this lies in the nature of the Mendelian mechanism. What is transmitted by a parent to his progeny is not an intact genotype, but a random sample of genes. Therefore the best prediction is a linear estimate of the average value of the contribution of the individual genes, the variance of which is measured by V_G. I might add, parenthetically, that the situation is not really as tidy as the above sentences may imply. It is difficult to identify the contribution of epistasis, particularly when one considers the complications of linkage between genes on the same chromosome. In many cases the breeder gets satisfactory predictions by simply ignoring epistasis, a fact which may be caused by one of several conditions. It may be that the gene loci act approximately additively on the chosen measurement; it may be that various gene interactions are in opposite directions and therefore cancel each other; or, as is often the case, it may be that the numbers are small, so that even a large discrepancy is regarded as a satisfactory fit, simply because there is not enough statistical power to make a finer discrimination.

The other interactions, between genotype and environment and between environmental components, are not automatically taken care of and have to be considered specifically. It is important conceptually to distinguish, as Jensen does, between *interaction* of heredity and environment (as when a good genotype gets more of a boost from a good environment than a poor genotype does) and *covariation* of heredity and environment (when a good genotype tends to be located where the environment is good). The components due to errors of measurement can usually be ignored if the correlations are corrected for attenuation.

It is important to emphasize that heritability can be defined in two ways:

$$\text{heritability in the narrow sense: } h^2 = \frac{V_G}{V_P}$$

$$\text{and heritability in the broad sense: } H^2 = \frac{V_H}{V_P}.$$

For mnemonic convenience, I shall use h^2 to indicate the narrow definition, which is always quantitatively smaller, and H^2 for the broader (and larger) definition.

Finally, I designate the environmental fraction of the variance as

$$E^2 = \frac{V_E}{V_P}$$

The plant or animal breeder is interested in h^2 because it helps him to predict the expected gains from selection and to estimate the effectiveness of a breeding program. The psychologist is likely to be more interested in H^2 (and E^2) because it partitions the variance into genotypic and environmental components and may thereby afford some insight into each cause. E^2 gives some guidance as to the amount of influence that environmental differences are having and, among these, specific factors may be identified. (See Jensen's discussion of heritability, pp. 33-43.)

Animal and plant experiments have shown that heritability estimates have reasonably good predictive accuracy when the numbers and statistical design are such as to provide a powerful test. However, the prediction is valid for only this particular situation, because heritability is a function of gene frequencies, the mating system, and existing environmental influences. As such, it will change when these change. This means, among other things, that the initial heritability will not be a good guide for long-time selection programs. The program, if successful, will change gene frequencies, and therefore the heritability may change. Furthermore, the environment may change, and this can also change the heritability.

How Valid Are Heritability Measurements of Intelligence?

The animal or plant geneticist gets rid of some of the most troublesome covariances by experimentally designed randomization. This is clearly out of the question in dealing with man. Correlations between relatives are caused by both genetic and environmental similarities. Jensen's general formula for H^2—a great improvement over those of earlier authors, in my opinion—is

$$H^2 = \frac{r_1 - r_2}{\rho_1 - \rho_2} ,$$

where r_1 and ρ_1 are the observed and theoretical correlations for one degree of relationship, and r_2 and ρ_2 are the corresponding quantities for another degree. If we ignore interaction, the correlation will be

$$r_i = \rho_i H^2 + \rho_i' E^2 ,$$

where ρ_i' is the environmental correlation for the ith degree of relationship. As a simple illustration of what happens when the ρ's are not independent, suppose that $\rho_i' = k\rho_i$, where k is a constant. Then

$$\frac{r_1 - r_2}{\rho_1 - \rho_2} = H^2 + kE^2 ,$$

which, instead of measuring H^2, includes an unknown fraction of the environmental variance, thus limiting the usefulness of such a formula. It is likely, for example, that cousins have environmental similarities that are less than those for siblings but more than for children in unrelated families.

If the formula is used to compare monozygous and dizygous twins, there are the often-discussed uncertainties as to whether intra-family environmental differences are the same for the two kinds of twins, as the formula assumes. Identical twins may more often share experiences than dizygotic twins. But, as many authors have pointed out, environmental similarity for monozygotic twins is not necessarily greater than for dizygotic, especially when intra-uterine environment is considered. For example, the likelihood of an unequal blood supply is greater in monozygous twins. Finally, the value of ρ is uncertain. For monozygotic twins it is clearly 1. But for dizygotic twins it is not known exactly. It is roughly 1/2—but decreased by dominance and epistasis and increased by assortative mating, both by unknown amounts.

Most of these difficulties could be removed if individuals of close genetic relationship could be randomized with respect to environments and if unrelated children could be reared in identical environments. These conditions are partially met by studies of twins and siblings reared in different households and by unrelated children reared in the same. As Jensen mentions, the Burt study appears closest to the ideal of placing the separated identical twins into random environments at an early age. There is some reassurance to the skeptic (such as I have been) in that H^2, as estimated by the correlation of one-egg twins reared apart, and E^2, estimated by the correlation between unrelated children reared together, add up to approximately 1—as they should if everything is simple (.75 + .24 = .99 in Jensen's Table 2; .86 + .25 = 1.11 in Burt's individual measurements). Other crosschecks are also in reasonable agreement, but the numbers are small.

If we take the results from many investigations at face value, there is a great deal of consistency, as Jensen points out, and H^2 averages about .8. Furthermore the dominance and epistatic components appear to be small. That the heritability is large is a justifiable conclusion at this stage, although the precise value must remain in doubt for the various reasons given. We shall have to be content with measures that are only approximate, pending more evidence on the reality of the assumptions. I agree with Jensen in deploring an uncritical assumption that only environmental factors are important and that genetic differences are negligible.

I admire the diligence of Newman, Shields, and Burt in finding twins and siblings reared apart. Particularly useful, as Jensen has emphasized, would be data on half-siblings reared in different homes. Any excess of similarity of those with a common mother over those with a common father would provide a measure of prenatal and neonatal maternal influences. Though difficult to obtain, it would also be necessary to have data on the correlations between the non-common parents. There will always be some doubt, however, as to whether children from broken homes—separated twins and siblings—and from foster home environments can be regarded as representative of the normal population.

How Important Is It to Measure Heritabilities?

I share Dr. Jensen's interest in trying to determine H^2 and h^2, especially if this information can be extended to other populations. Particularly interesting is his suggestion that heritability be used as one criterion of the culture-fairness of a test. At the same time there are many social decisions that do not depend on a precise knowledge of the heritability of intelligence.

If society decides to improve IQ by eugenic means, h^2 will be useful in providing estimates of the expected gain. I believe that we already know enough to predict that a selection program to increase IQ or g would work. There would be an increase, but the amount would be uncertain, because of uncertainties both in the true value of h^2 and in the asumptions underlying its use as a predictor. However, society is clearly not ready to embark on a eugenic program of sufficient scope to make very much difference, even if heritability were equal to 1.

What guidance does H^2 (or E^2) offer for predicting the effect of improvement in the environment? E^2 tells us how much the variance would be reduced if the environment were held constant. It does not directly tell us how much improvement in IQ to expect from a given change in the environment. In particular, it offers no guidance as to the consequences of a new kind of environmental influence. For example, conventional heritability measures for height show a value of nearly 1. Yet, because of unidentified environmental influences, the mean height in the United States and in Japan has risen by a spectacular amount. Another kind of illustration is provided by the discovery of a cure for a hereditary disease. In such cases, any information on prior heritability may become irrelevant. Furthermore, heritability predictions are less dependable at the tails of the distribution.

A high heritability of intelligence does not necessarily mean that a program of compensatory education is destined to fail, although it may necessitate a larger or more innovative environmental change than if E^2 were larger. Measuring heritability may be less important than getting empirical data on the effects of specific environmental factors. If environment acts as a threshold, as Dr. Jensen suggests, then it would be especially important to identify environmental influences that may be of great influence at the end of the scale, but less so within the normal range.

I am not acquainted with the compensatory education studies Jensen reviews nor am I professionally competent in that area, but my view from the outside is that we should not give up too easily. Perhaps the programs are too little and too late. There are surely a variety of ways of intensifying and improving the effectiveness of education. Also a small change in IQ, especially if accompanied by increased motivation and achievement, may be of great social benefit. Jensen expresses much the same idea:

Thus it seems likely that if compensatory education programs are to have a beneficial effect on achievement, it will be through this influence on motivation, values, and other environmentally conditioned habits that play an important part in scholastic achievement, rather than through any marked direct influence on intelligence *per se*. The proper evaluation of such programs should therefore be sought in their effects on actual scholastic performance rather than in how much they raise the child's IQ. (p. 59)

Group Differences, Especially Racial Differences

Heritability studies have been confined almost exclusively to white populations and largely to normal environments. How relevant are they to other populations and environments? We are currently especially concerned about culturally disadvantaged groups and racial minorities. Strictly, as Jensen mentions, there is no carryover from within-population studies to between-population conclusions.

I agree that it is foolish to deny the possibility of significant genetic differences between races. Since races are characterized by different gene frequencies, there is no reason to think that genes for behavioral traits are different in this regard. But this is not to say that the magnitude and direction of genetic racial differences are predictable.

It is clear, I think, that a high heritability of intelligence in the white population would not, even if there were similar evidence in the black population, tell us that the differences between the groups are genetic. No matter how high the heritability (unless it is 1), there is no assurance that a sufficiently great environmental difference does not account for the difference in the two means, especially when one considers that the environmental factors may differ qualitatively in the two groups. So, I think, evidence regarding the importance of heredity in determining group mean differences must come from other kinds of studies.

The failure, thus far, to find identifiable variables that, when matched, will equalize the IQ scores does not prove that the mean difference is hereditary. It can be argued that being white or being black in our society changes one or more aspects of the environment so importantly as to account for the difference. For example, the argument that American Indians score higher than Negroes in IQ tests—despite being lower on certain socio-economic scales—can and will be dismissed on the same grounds: some environmental variable associated with being black is not included in the environmental rating. Behavioral scientists can be expected to disagree, and they do, as to when enough identifiable environmental factors have been shown to be insufficient that the remaining differences should be regarded as mainly genetic. To me, the evidence on this question is not at all conclusive.

Final Comments

One of the goals of a democratic society, I believe, ought to be to provide each individual with the maximum opportunity to satisfy his needs and desires and to contribute to society's betterment through his special abilities. A population with a variety of phenotypes (and genotypes) ought to be more rewarding, and certainly more interesting, than one that is homogeneous. I do not go to the extreme of saying that all variation should be encouraged—I shall be quite happy if the gene for muscular dystrophy becomes extinct—but in general I believe that diversity is good, not bad. In any case, we have it.

Society should recognize that there is a great deal of genetic variability for all kinds of traits, including intelligence and special talents. I think that J. B. S.

Haldane once said that liberty is the practical recognition of human variability. We should also realize that to whatever extent society is successful in its goals of providing equality of opportunity, to that extent the heritability will increase. In view of this fact, I fully agree with Jensen that, rather than uniformity, the goal should be diversity of educational opportunity with maximum individual opportunity for finding the right niche, and that the reality of individual differences need not and should not mean rewards for some and frustration for others.

The Future of Individual Differences

CARL BEREITER, *The Ontario Institute for Studies in Education*

Professor Bereiter concurs in Jensen's re-emphasis of the heritability of intelligence, but he draws different conclusions about the probable future. Because most intellectual tools which can be learned act as amplifiers rather than equalizers of basic differences in problem-solving ability and because our complex society increasingly emphasizes intelligence rather than other abilities, Bereiter believes that the kind of educational effort recommended by Jensen may in fact only increase the consequences of individual differences. Nevertheless, he suggests that this pessimistic projection may be open to revision in the light of ongoing work in early remedial education.

I have read Dr. Jensen's paper as an essay on the subject of what lies beyond attainment of equal educational opportunity. He does not deny that educational inequities exist and should be rectified, but he is concerned that people seem to expect the removal of such inequities to eliminate effectively the great spread of individual differences in intelligence, with its host of social concomitants. Dr. Jensen expects that in reality the removal of these inequities will have little effect on the spread of individual differences and he proposes that we start developing educational approaches that accept this spread of abilities as hard fact.

My own view of the future of individual differences and their social consequences is even less optimistic than Dr. Jensen's. The heritability of intelligence is unquestionably high, but what is more to the point is that with further social progress its heritability can only increase, because of the elimination of such sources of environmental variance as differences in the quality of education, nutrition, and medical care. One's view of the future beyond equality of opportunity

Harvard Educational Review Vol. 39 No. 2 Spring 1969, 310–318

must, therefore, be of a future in which differences in intelligence are virtually one hundred percent determined by heredity.[1]

The magnitude of these differences then becomes a crucial question; however, it is not magnitude in terms of IQ points that counts, but magnitude in terms of differences in effective problem-solving capability.[2] These are not the same thing, even if they are perfectly correlated. We may expect that through continued scientific progress, through the continued development of intellectual "tools" such as language, logic, thinking machines, and scientific techniques, man's ability to solve problems will increase at an accelerating pace, even though his IQ changes not a jot. There is nothing paradoxical in this. Through the development of mechanical tools man's ability to lift weights, hurl objects, and so on, has been multiplied manyfold, independently of any increase in his basic muscular strength.

Tools, then, may act as amplifiers or equalizers with respect to basic human capabilities. A lever, for instance, amplifies force. If it triples the force Smith and Jones can exert, it also triples the *difference* between the forces Smith can exert and Jones can exert. An electric hoist, on the other hand, is an equalizer. So long as they both have what it takes to operate the hoist, both Smith and Jones can lift the same weight, regardless of their differences in muscular strength. Technology has generally been moving toward tools of the latter type, thus giving rise to the spectre of a future automated world in which individual human differences will no longer count for anything, having been obliterated by the uniformity of machine performance. I would edit this vision in only one important respect: in this future world the overwhelming variable of individual differentiation will be that of intelligence as manifested in the ability to use those tools that make other individual differences irrelevant. This statement rests on the following arguments:

1. The equalizing effect of sophisticated tools is gained by having intelligence take over the function of other abilities. Whereas the photographer once needed the ability to judge depths and levels of illumination accurately in order to take a clear picture, he can now be quite deficient in these abilities providing he is intelligent enough to use his equipment properly. This is not to say that sophisticated tools always require more intelligence to operate than primitive ones;

[1] This eventuality is in no wise to be forestalled by individualized instruction or any more libertarian tactic; on the contrary, such approaches should allow inherited differences to reach full flower, as advertised in the slogan, "enabling each child to realize his fullest potential."

[2] Another way of saying this is that it is surface traits rather than source traits of intellectual ability that count socially (Cattell, 1950). Both, of course, are phenotypes.

163

they may require less. The point is that the sophisticated tool requires only intellectual ability, whereas the primitive tool required intelligence plus physical strength, manual dexterity, sensory acuity, etc.

2. Intellectual tools, by which I mean algorithms, principles, systems, and devices that are used in processing information, appear in the long run always to function as amplifiers rather than equalizers of intelligence and, thus, to magnify rather than nullify individual differences in ability. The class of tools called mathematics furnishes the clearest examples. Using only arithmetic, people will differ considerably in the complexity of problems they can solve with it, as shown for instance by their performance on the Arithmetic Reasoning Test of the Wechsler Adult Intelligence Scale. Using elementary algebra, the less capable person will be able to accomplish little more than he could with arithmetic, whereas the more capable will be able to use it to solve problems of quite a high level. Using analytical geometry and calculus, the less capable person will again show little gain, except for being able to solve integration and differentiation problems that are clearly set up for him, whereas the more capable person will now be able to solve problems that the duller one cannot even conceive of.

Even when a new tool serves as an equalizer with respect to ability to solve a certain kind of problem, its overall effect seems to be that of an amplifier. Computer programs for the rotation of factors in factor analysis and for the simplification of electrical circuit designs have taken over tasks that used to require considerable art and intelligence. But the ultimate effect of such a development is simply to take out of the hands of specialists and to make available for more general use tools which the intelligent person can use intelligently and the unintelligent person can use unintelligently, thus increasing their manifest difference.

3. Every tool requires certain minimum abilities of a person in order for him to use it at all. Accordingly, each new tool drives a wedge between those who can learn to use it and those who cannot. The more powerful the tool the wider the wedge. An enormous effective gap, for instance, separates those who can learn to read from those who cannot. The social importance of such gaps seems, however, to depend not only on their size, but on where they occur in the distribution of abilities—on whether they separate a small minority at the top or bottom of the distribution or whether they separate the population more nearly into halves.

The direction of progress in the development of intellectual tools and of methods for teaching their use is generally toward lowering the level of intellectual ability prerequisite to their use. Thus, I do not see Dr. Jensen's proposal, that educators look for ways to make school learning less dependent on intelligence, as a very radical one. This is what efforts at curriculum improvement, remediation, and improvement of teaching methods all try to do, whether successfully or not.[3] Bringing intellectual tools within the reach of more people does not necessarily have a leveling effect, however. To replace a method of reading instruction in which fifteen percent of children fail by one in which only five percent fail would tend to make for more social equality; but to take some powerful intellectual tool that could be mastered by only one person in a hundred and to make it so that half the population could master it would be a divisive influence. One such possibility is suggested by the oft-mentioned prospect of household computers. Presumably, such computers would be so simple in their routine operation that all but the most incompetent could manage them. But being general purpose computers, they would lend themselves to all sorts of non-routine applications and thus would provide an intellectual tool of enormous power and versatility to the person who could program them—and this is an ability that might well be put within the reach of about half the population, and yet remain unattainable by the other half. Thus, a sharp and conspicuous split in effective problem-solving ability would arise where none exist at this time.

The magnitude of individual differences referred to in the above arguments is, of course, a subjective matter, a matter of what people make of perceived differences rather than of objective magnitude. There is no absolute sense in which one could say that individual differences in problem-solving ability are greater than, say, individual differences in perceptual abilities; yet as a statement about the recognized and pragmatically significant differences among people in modern societies, it is obviously valid. We are sharply aware of differences among our fellow men in problem-solving ability; such perceived differences figure in countless decisions, with the result that a person's problem-solving ability enters prominently and complexly into the determination of his social fate. Differences in perceptual abilities, on the other hand, come to our attention only rarely and in special circumstances and for most people play little part in determining the course and character of their social lives. It is easy to imagine a world in which the tables would be turned.

---

[3] Even when new mathematics and science curricula demand more intelligence of the students than old curricula, if one considers what the new curricula are actually trying to teach, it will be seen that they are trying to bring within the reach of a wider population concepts and tools that were previously reserved for a more advanced or gifted minority.

Dr. Jensen has noted that there are societies in which differences in g do not count for so much. Presumably, there is a level of description at which it may be said that individual differences in intelligence are of about the same magnitude in those societies as they are in ours, but the level of description is assuredly not that of manifest effectiveness in solving real-life problems or of differentiation of social status on the basis of competence. A common interpretation of this anthropological fact is that, for one or another creditable reason, some other societies do not "value" intellectual abilities as we do. Another interpretation is that these societies lack the intellectual tools that would amplify differences in problem-solving abilities to the point where they are as conspicuous as they are in our society.

Either way, as one moves from relatively primitive to relatively advanced societies, individual differences in intelligence become at once more conspicuous and more consequential in manifold ways. What I have been proposing is simply that this trend will continue into the future at a rapidly accelerating rate, as differences in intelligence take the place of more and more other, formerly compensating, differences in ability and as more and increasingly powerful intellectual tools become available to magnify differences in effective intellectual ability.

We may now hasten to the denouement of this pessimistic story. The prospect is of a meritocratic caste system, based not on arbitrary distinctions of privilege as in traditional caste systems, but on the natural consequences of inherited differences in intellectual potential. These consequences, however, could be expected to extend well beyond differences in occupational status, to include associated differences in attitudes, interests, and ways of life. Assortative mating could be expected to intensify under these conditions, thus leading to further augmentation of inherited differences and rigidification of the caste hierarchy. Such a caste system would be far more resistant to democratizing influences than imposed caste systems of the past. It would tend to persist even though everyone at all levels of the hierarchy considered it a bad thing. The already high level of assortative mating on intelligence, which according to Dr. Jensen is higher than on any other trait that has been investigated, is perhaps the strongest single piece of evidence that progress toward this caste system is already well advanced.[4]

[4] Michael Young's otherwise compelling fictional account of *The Rise of the Meritocracy* (1958) misses the mark, I believe, in focussing on the tyrannical use of IQ tests to fix people's places in the meritocratic hierarchy. Testing is a red herring in this discussion, for it could at best be used only to facilitate discriminations that would be made anyway. The great improvements in intelligence testing that Young saw as necessary to the fullest development of meritocracy are not only unnecessary but also unlikely. The validity of intelligence tests has not increased appreciably in thirty years, and there is little prospect that they will ever account for more than about half the variance in non-test criteria of achievement. Improvements in instructional tech-

In this futuristic context, and in light of the failure of education to date, it may seem gratuitous to raise the question of whether education can do anything to equalize effective intelligence (once it has accomplished the still far from realized goal of giving everyone equal access to intellectual tools). I am encouraged to keep the question open, however, if only because of the results of early education experiments that I have had a part in. The approach that I and my co-workers have taken to early education of disadvantaged children has been rather close to that which Dr. Jensen advocates. We were not trying to "stimulate the growth of intelligence," but rather to teach academic skills directly in ways that did not demand of the children abilities they demonstrably did not possess (Bereiter, Engelmann, Osborn, and Reidford, 1966; Bereiter and Engelmann, 1966). As judged by achievement tests, the efforts have been quite successful (Bereiter, 1968), and I think they lend support, at least as far as the early stages of subject-matter learning are concerned, to Jensen's conviction that "all the basic scholastic skills can be learned by children with normal Level I learning ability, provided the instructional techniques do not make g ... the *sine qua non* of being able to learn" (p. 117).

Nevertheless, in spite of the fact that the program was never intended to raise IQ and that two-thirds of it was devoted to reading and arithmetic instruction having little or nothing to do with the skills called for on IQ tests, significant IQ gains have been regularly obtained. Over the last four replications they have averaged about 15 points. This seems to be too much of a gain to write off to test-wiseness and things of that sort, especially since the children's IQs were in the middle nineties to begin with and thus rose to substantially above average.

However, we never entertained any illusions that the instruction was improving the children's brains. The most reasonable interpretation had seemed to be that the IQ gains merely reflected the accelerated learning of some kinds of conceptual content sampled by the IQ test (in all cases the Stanford-Binet). This interpretation has received something of a blow, however, from a recent and as yet unpublished study in which we tried out a new curriculum generated by working backward from the Stanford-Binet to create a universe of content for

nology, which would make instruction continuously adaptive to variations in level, rate, and style of learning, are foreseeable, however, and would render IQ testing irrelevant. They would also have the effect of streaming people into different levels of the meritocratic hierarchy without the least hint of coercion. By imagining a tyrannical system, Young imagined one that could be overthrown by the oppressed. The currently high level of assortative mating on intelligence demonstrates how little meritocracy need depend either on IQ-branding or on official control.

which the Stanford-Binet could be considered a content-valid achievement measure. Going at it in this bald-faced manner, we expected to obtain enormous but, of course, psychologically meaningless IQ gains on the Stanford-Binet. As a check on non-specific effects, we also used the WPPSI as a pre- and post-test, without its content's being known during the experiment either to the curriculum writers or to the teachers. Contrary to expectation, the gains on the Stanford-Binet were not large compared to those regularly obtained with the academically-oriented curriculum—about 12 points, and the gains on the WPPSI were exactly the same as those on the Stanford-Binet.

Tracking down what is actually learned in order to account for IQ gains is likely to prove an arduous and perhaps ultimately thankless task. Our unpublished study does not point to any answer but does suggest that there may be more to educationally-induced IQ gains than meets the eye, whereas we, along with Dr. Jensen, had been inclined to assume that there was less.

Here is one possibility: thinking, as even the behaviorists are coming to admit, must surely consist of very long strings of actions or responses, most of which are never directly subjected to corrective feedback. Thus learning to think, to the extent that it occurs, must occur under less than propitious circumstances. This would constitute a situation in which inherited differences in functioning of an otherwise trivial nature could have profound effects. Slight differences in immediate memory, alertness, etc., could spell the difference between learning and not learning some of the cognitive behaviors involved in thinking, or between learning them early and learning them late. Yet with a little help they might be learned—help of a kind that is not regularly provided by the feedback conditions of either normal or school life. Educational programs that produce substantial IQ gains may have inadvertently managed to teach such behaviors. The temporary character of IQ gains doesn't negate this possibility. Most likely such programs, operating blindly in this regard, could do no more than teach early what would be learned later anyway, so that IQs eventually return to their expected levels.[5]

---

[5] Whether this will be altogether the case with children educated in our program remains to be seen. It was so for the children in the original pilot group. By the end of second grade their IQs had gone back down to their original level. However, the mean IQ for that group had only risen 10 points after treatment. The second wave, on the other hand, showed a 25 point IQ gain in two years of preschool treatment. By the end of first grade their IQs had declined 11 points, but this still left them with a net gain of 14 points and a mean IQ of 110. A randomly equivalent control group given one year of Head Start-like enrichment gained approximately 8 points and remained at the end of first grade with a net gain of 5 points and a mean IQ of 101 (Karnes, 1969).

In order to achieve any lasting neutralization of the inherited tendencies leading to lower IQ, it would be necessary to discover cognitive behaviors which duller people will never learn at all and to find ways of teaching them. I will not pretend to specify any such behaviors, even speculatively, but I will suggest a couple of areas where they might lie. One is an area that may be called preliminary information-processing—what you do with incoming information when you don't yet have enough other information to make intelligent use of it. I would suggest that the dull person doesn't do anything with it most of the time, so that he is said to have an attentional deficit (Zeaman and House, 1963), whereas the intelligent person has learned a number of provisional information-processing moves which at least have the effect of preserving the pieces of information in a form so that they can be assembled later (Payne, Krathwohl, and Gordon, 1967). Another is in the construction of solution models for problems (Gagne, 1966), which even some college students seem to do not merely poorly, but not at all (Bloom and Broder, 1950).

Remedial education, along with remedial genetics and remedial biochemistry, might conceivably have some appreciable effect in reducing the spread of individual differences in intelligence. I see no prospect whatever, however, for a reversal of the tendency for intelligence to take over the function of other human abilities. That tendency is intrinsic in the entire progress of science and technology. The domain in which other human abilities are significant becomes increasingly limited to sports and to arts where the scope of intelligence is arbitrarily restricted (through restrictions on the kind of equipment that may be used, for instance). Thus Dr. Jensen's closing appeal for diversity of aims in education inspires more nostalgia than hope, recalling the nearly vanished era when blacksmith, watchmaker, woodcarver, gardener, and a host of others could attain some measure of distinction on the basis of special abilities little related to general intelligence. Special abilities will continue to have a place, of course, but as adjuncts rather than alternatives to general intelligence.[6] If we are to

[6] Throughout this discussion I have followed Dr. Jensen in using the terms g, intelligence, and IQ interchangeably. I don't believe that either his argument or mine would be materially altered by dropping the notion of g and adopting a multifactorial view of intellectual abilities, as in Guilford (1967). The main difficulty would be the shortage of relevant data on separate intellectual abilities, compared to what is available on general intelligence. A special drawback to approaching the problem multifactorially is the lack of data on hereditary and environmental contributions to the *correlations* between mental abilities. According to Thompson (1966), this matter has never been studied, although the methodology is available and has been applied to other problems. If we regard g as an unrotated first factor (Rimoldi, 1951), its composition would naturally change with change in the selection of tests, as Dr. Jensen notes, and preferred

make something of the "untapped reservoir" of learning ability that Dr. Jensen finds among the disadvantaged, it would seem that we must look—as educators and psychologists have really only just begun to do—for ways to marshal this learning ability to the task of learning to think.

---

selections of tests might change as cultural conditions changed. For instance, there could be a shift toward greater emphasis on creativity measures. Such shifts would have implications for who ranks where in a meritocratic hierarchy but not, foreseeably, of such a radical kind as to require serious qualifications in any arguments presented here.

References

Bereiter, C. A non-psychological approach to compensatory education. In M. Deutsch, I. Katz, and A. R. Jensen (Eds.), *Social class, race, and psychological development.* New York: Holt, Rinehart & Winston, 1968.

Bereiter, C. and Engelmann, S. *Teaching disadvantaged children in the preschool.* Englewood Ciffs, N. J.: Prentice-Hall, 1966.

Bereiter, C., Engelmann, S., Osborn, J., and Reidford, P. An academically-oriented pre-school for disadvantaged children. In F. M. Hechinger (Ed.), *Pre-school education today.* Garden Hills, New York: Doubleday, 1966, pp. 105-135.

Bloom, B. S. and Broder, L. J. *Problem-solving processes of college students.* Chicago: University of Chicago Press, 1950.

Cattell, R. B. *Personality: A systematical theoretical and factual study.* New York: McGraw-Hill, 1950.

Gagne, R. M. Human problem solving: internal and external events. In B. Kleinmuntz (Ed.), *Problem solving: Research, method, and theory.* New York: Wiley, 1966.

Guilford, J. P. *The nature of human intelligence.* New York: McGraw-Hill, 1967.

Karnes, M. B., Teska, J. A., Hodgins, A. S. A longitudinal study of disadvantaged children who participated in three different preschool programs: Traditional, direct verbal, and amelioration of learning deficits. Paper read at American Education Research Association, Los Angeles, February, 1969.

Payne, D. A., Krathwohl, D. R., and Gordon, J. The effect of sequence on programmed instruction. *American Educational Research Journal,* 1967, 4, pp. 125-132.

Rimoldi, H. J. A. The central intellective factor. *Psychometrika,* 1951, 16, pp. 75-102.

Thompson, W. R., Multivariate experiment in behavior genetics. In R. B. Cattell (Ed.), *Handbook of Multivariate Experimental Psychology.* Chicago: Rand-McNally, 1966, pp. 711-731.

Young, M. *The rise of the meritocracy.* London: Thames & Hudson, 1958.

Zeaman, D., and House, B. J. The role of attention in retardate discrimination learning. In N. Ellis (Ed.), *Handbook on Mental Deficiency.* New York: McGraw-Hill, 1963.

Piagetian and Psychometric
Conceptions of Intelligence

DAVID ELKIND, *University of Rochester*

Professor Elkind devotes much of his discussion to the concept of intelligence. He finds both similarities and differences when comparing the Piagetian description of intelligence with Jensen's (and the psychometrician's) definition of intelligence. Operating from quite different assumptions than those of J. McV. Hunt (Piaget's Structuralism, rather than neurology) Elkind also finds reason to believe that intelligence is developed in experience. For Piaget and Elkind, intelligence is "an extension of biological adaptation" and is characterized by ability to assimilate (develop in response to internal processes) and accommodate (respond to environmental intrusions).

I have been asked to respond to Professor Jensen's paper from the standpoint of Piaget's genetic psychology of intelligence. While I clearly cannot speak for Piaget, only the "Patron" can do that, I can react as someone steeped in Piagetian theory and research and as one who looks at cognitive problems from the Genevan perspective. Accordingly, while I hope that what I have to say would be acceptable to Piaget, I cannot guarantee that this is in fact the case, and must take full responsibility for whatever is said below. I plan to discuss, in the first section of the paper, some of the similarities between the Piagetian and psychometric positions. Then, in the second section, some of their differences will be pointed out. Finally, in the third section, I want to consider two related practical issues regarding the modification of intelligence.

Harvard Educational Review Vol. 39 No. 2 Spring 1969, 319-337

171

Conceptual Similarities

What struck me in reading Professor Jensen's paper, and what had not really occurred to me before, were the many parallels and affinities between the psychometric or mental test approach to the problem of intelligence and the developmental approach as represented by Piaget. It brought to mind the fact that Piaget began his career as a developmental psychologist by working in Binet's laboratory where he sought to standardize some of Burt's (1962) reasoning tests on Parisian children. Indeed, Piaget's *method clinique* is a combination of mental test and clinical interview procedures which consists in the use of a standardized situation as a starting point for a flexible interrogation. The affinities, however, between the Piagetian and psychometric approaches to intelligence run more deeply than that. In this section I want to discuss such affinities: the acceptance of genetic and maturational determination in intelligence, the use of non-experimental methodologies and the conception of intelligence as being essentially rational.

Genetic Determination

Implicit and often explicit in both the psychometric and Piagetian positions is the assumption that mental ability is, in part at least, genetically determined. With respect to the psychometric position, it assumes that at least some of the variance in intelligence test performance is attributable to variance in genetic endowment (Burt & Howard, 1957, Jensen). Piaget (1967a) also acknowledges the importance of genetic factors for intellectual ability but qualifies this by pointing out that what may be genetic in one generation may not always have been so and could be the partial result of prior environmental influences. So, for Piaget, as for the biologist Waddington (1962a) there is a certain relativity with respect to what is attributed to genetic endowment because what is genetic now may not always have been genetic. To illustrate, Waddington (1962a) observed that after several generations a strain of the fly grub drosophilia developed enlarged anal papillae when reared on a high salt diet. When the insects were returned to a "normal" low salt diet the anal papillae of successive generations became less large but never returned to their original size. Waddington speaks of this as "genetic assimilation" by which he means that the effects of an altered environment upon the selection process within a species may not be completely reversible even when the environment returned to its unaltered state.

One consequence of their joint acceptance of the partial genetic determination of intellectual ability, is that both psychometricians and Piaget recognize the importance of maturation in human development. To illustrate their commonality in this regard, consider these two passages, one written by Harold Jones in 1954 and the other by Piaget in 1967.

Dubnoff's work, together with other related studies, may lead to the speculative suggestion that between natio-racial groups, as within a given group, a slight tendency exists for early precocity to be associated with a slower mental growth at later ages and perhaps with a lower average intelligence level at maturity. A parallel situation may be noted when we compare different animal species; among the primates, for example, the maturity of performance at a given age in infancy can be used inversely to predict the general level of adaptive ability that will be attained at the end of the growth span. (Jones, 1954, p. 638)

And Piaget writes:

We know that it takes 9 to 12 months before babies develop the notion that an object is still there when a screen is placed in front of it. Now kittens go through the same substages but they do it in three months—so they are six months ahead of the babies. Is this an advantage or isn't it? We can certainly see our answer in one sense. The kitten is not going to go much further. The child has taken longer, but he is capable of going further so it seems to me the nine months were not for nothing. (Piaget, 1967b)

Non-Experimental Methodology

In addition to their shared genetic or maturational emphasis, the Piagetian and psychometric approaches to intelligence have still another characteristic in common. This common feature is their failure, for the most part, to use the experimental method in the strict sense of that term. It seems fair to say that most of the studies which attempt to get at the determinants of test intelligence are correlational in nature. By and large such studies attempt to relate the test scores of parents and their children, of twins or of adopted children and their parents, or of the same children tested at different points in time and so on. Only in rare instances such as the Skeels (1966) study is an attempt made to modify intelligence by active intervention and with the utilization of a control group which does not receive the experimental treatment. While experimental work on human intelligence might well be desirable, such research often raises serious moral and ethical questions.

Piaget, for his part, has not employed the experimental method simply because it was not appropriate for the problems he wished to study. This is true because Piaget has been primarily concerned with the diagnosis of mental contents and abilities and not with their modification. To illustrate, the discovery of what the child means by "more," "less" and "same" number of things requires flexible diagnostic interview procedures and not experimental procedures. Once the concept is diagnosed, then experimental methods are appropriate to determine the effects of various factors on the attainment and modification of the concepts in question. The sequence of events is not unlike the situation in medicine where the discovery or diagnosis of a disease is often the first step to its experimental investigation. In short, Piaget has focused upon the discovery of what and how children think and not with the modification of thinking which is a subsequent and experimental question. In every science there is a natural history stage of enquiry during which relevant phenomena must be carefully observed and classified. American psychology has often tried to bypass this stage in its headlong rush to become an experimental science. In his studies Piaget has revealed a wide range of hitherto unknown and unsuspected facts about children's thinking, which have in America now become the starting points for a great deal of experimental investigation. What is often forgotten, when Piaget is criticized for not using the experimental method, is that such a method would not have revealed the wealth of phenomena which experimental investigators are now so busily studying.

Rationality as the Definition of Intelligence

There is a third and final commonality in the mental test and Piagetian approaches to intelligence which should be mentioned. This commonality resides in what these two positions regard as the nature or essence of intelligence. While there is considerable variability among psychometricians in this regard, many agree in general with the position taken by Jensen (1969). Jensen argues that the g factor which is present in all tests of mental ability appears in its purest forms on tests of generalization and abstraction. Spearman (1923) called these activities the education of relations (A is greater than B; B is greater than C; so A is in what relation to C?) and of correlates (Complete the series A AB ABC ———). While intelligence tests contain measures of many different types of mental abilities, including language and perceptual skills, the psychometric approach holds that the most central feature of human intelligence is its rationality, or as Wechsler put it: "Intelligence is the aggregate or global capacity of the individual to act purposefully, to think rationally and to deal effectively with his environment" (Wechsler, 1944, p. 3).

For Piaget, too, the essence of intelligence lies in the individual's reasoning capacities. Piaget, however, is more specific in his description of these abilities and defines them in terms of mental operations which have the properties of mathematical groupings in general and the property of reversibility in particular. An operational grouping is present when in the course of any mental activity one can always get back to the starting point. For example, if the class *boys* and the class *girls* is mentally combined to form the class *children*, it is always possible to recapture the subclass by subtraction. That is to say, the class of children minus the class of boys equals the class of girls. Put differently, the operation of subtraction can be used to undo the operation of addition so that each of the combined classes can be retrieved. Verbal material learned by heart is, however, not rationally organized as is illustrated by the fact that no matter how well a passage is learned, it is impossible, without additional effort, to say it backwards. If an operational system were involved, having learned the passage forward would automatically imply the ability to say it backwards. In Piaget's view, neither perception nor language are truly rational since neither one shows complete reversibility. So, while perception and language play an important part in intellectual activity, they do not epitomize that activity.

The psychometric and Piagetian approaches to intelligence thus agree on its genetic determination (at least in part), and on the use of non-experimental methodology and upon the essentially rational nature of mental ability. After this look at their commonalities, it is perhaps time to look at their differences.

Conceptual Differences

Despite the commonalities noted above, the psychometric and developmental approaches to intelligence also differ in certain respects. These differences, however, derive from the unique ways in which the psychometricians and Paiget approach and view intelligence and not from any fundamental disagreements regarding the nature of intelligence itself. In other words the differences are due to the fact that the two approaches are interested in assessing and describing different facets of intelligent behavior. Accordingly the differences arise with respect to: (a) the type of genetic causality they presuppose; (b) the description of mental growth they provide; and (c) the contributions of nature and nurture which they assess.

Genetic Causality

Although the Piagetian and psychometric approaches to intelligence agree on the importance of genetic determination, at least in part, of human mental ability, each approach emphasizes a somewhat different mode of genetic determination or causality. In order to make these differences clear, it is necessary to recall some of the basic features of evolutionary theory upon which all modern conceptions of intelligence are based.

Within the Darwinian conception of evolution there are two major phenomena that have to be taken into account: within-species variability and natural selection. For any given species of animal or plant one can observe a range of variations in such features as color, shape and size. Among a flock of robins, to illustrate, one can see that some adult birds differ in size, in richness of breast coloration and that some even manifest slight variations in head and wing conformation. Similar variations can be observed among a group of collies, Persian cats and even among tomato plants in the garden. This within-species variability, we know today, is due to the chance pairings of parental genes and to gene complexes which occur because each parent contributes only half of its genetic complement to its offspring. Variations within a given species at a given time are, therefore, primarily due to chance factors: namely the random genetic assortments provided by the parent generation. One determinant of variability among animals and plants is then, simply, chance.

Now in the psychometric conception of intelligence, this random type of variation is just what is presupposed. Test intelligence, it is assumed, is randomly distributed in a given population at a given time and such distributions should resemble the bell shaped curve of the normal probability function. Measurement of human abilities does in fact reveal a tendency for such measurements to fall into normal distributions. In addition evidence such as "regression toward the mean" (children of exceptionally bright or dull parents tend to be less bright and less dull than their parents) is also characteristic of genetic traits which are randomly determined. In short, when the psychometrician speaks of genetic determination, he is speaking of the chance gene combinations which produce a "normal" bell-shaped distribution of abilities within a given population.

Obviously this description of genetic determination is extremely over-simplified; we know that a test score is a phenotype which is determined by many different factors not all of which are genetic. Jensen, to illustrate, breaks down the variance of test intelligence into a large number of components such as genotypic variation, environment, environment genotype interaction, epistasis, error of measurement variance and so on. With the exception, perhaps, of the selective mating variable, however, all of these factors can again be assumed to operate in a random manner so that one might say that the chance distribution of observed test scores is the product of many underlying chance distributions. That the psychometric approach does in general presuppose a random distribution is also shown by the fact that the criterion of a true change in intellectual ability is the demonstration that such a change could *not* be attributed to chance factors.

That variability within a species is in part determined by chance gene and gene complex assortments has of course been demonstrated by Mendel and all of the research which has derived from his theory of genetics. There are, however, other forms of organismic variability which cannot be attributed to chance. Natural selection, the other component of evolution, is never random but always moves in the direction of improved adaptation to the milieu. To illustrate, over the past hundred years there has been a gradual predominance of dark over light colored moths in the industrial sections of England. Kettlewell (1955) demonstrated the survival value of dark coloration by showing that light moths placed on soot darkened bark were more readily eaten by insectivorous birds than were similarly placed dark moths. When variations across generations are considered, the variations are not random but rather show a clear cut direction.

The same holds true within the course of individual development. In the case of individual growth, however, the direction of progress is not determined by mating practices but rather by biochemical mechanisms which are only now in the process of being understood. That these biochemical agents determine the direction of development, however, cannot be doubted. As Waddington (1962b) points out, animals consist of a limited variety of cells such as nerve cells, muscle cells and so on. Likewise the organs of the body are also distinct from one another in form, composition and function. What direction particular cells will take as the egg matures will depend upon the action of chemical agents which Spemann (discussed in Bertalaffny, 1962) called *organizers* with definite loci in the cell material called *organization centers*. It is the organizer which determines whether particular cells will become nerve, muscle or organ tissue. Individual development, therefore, is not determined by random factors but rather by biochemical organizers which specify the nature and direction of organismic differentiation.

Now when Piaget speaks of the genetic determination of intelligence, he has in mind not the random factors which determine gene combinations, but rather the non-random action of biochemical organizers and organization centers. Indeed, this is the kind of determination which Piaget assumes when he argues that the *sequence* in which the child attains the successive components of a concept or in which he acquires systems of mental operations, is invariant. In the formation of body organs the order of differentiation is fixed because each new phase of differentiation produces the organizer for the next stage. In Piaget's view this is equally valid for the growth of cognitive structures because the preceding cognitive structures, say the concrete operations of childhood, are a necessary prerequisite to the elaboration of the more complex formal operational structures of adolescence. For Piaget, then, genetic determination means that there are factors which give development a definite non-random direction.

In pointing out that the Piagetian and psychometric approaches to intelligence postulate different forms of genetic determinism, I want to reiterate that these two positions are not in contradiction one with the other. The mental test approach to intelligence is concerned with inter-individual differences in ability and these are, in so far as we know, largely randomly determined. Piaget, in contrast, is concerned with the intra-individual changes which occur in the course of development and these, to the best of our knowledge, are not random but rather have a direction given them by specific organizing mechanisms. Accordingly, and this is the genius of evolution, human intelligence manifests both determinism *and* freedom.

The Course of Mental Growth

Let us look now at a somewhat different issue, the age-wise course of mental growth. Here again we find a difference in perspective rather than a contradiction in conception as between the two positions. In psychometric terms, the course of mental growth is plotted as a curve which measures the amount of intelligence at some criterion age that can be predicted at any preceding age. As Bloom (1964) has pointed out, when age 17 is taken as the criterion age, some 50% of the total IQ at that age can be predicted at age four, and an additional 30% can be predicted from ages four to eight. Based on correlational data of this sort, curves of mental growth appear to rise rapidly in early childhood and taper off to a plateau in late adolescence. Such curves, it must be noted to avoid a frequent misinterpretation, say nothing as to the *amount* or *quality* of knowledge at given age levels. (See Jensen, pp. 115-117.)

From the mental test perspective, therefore, intellectual growth is pretty much a statistical concept derived from correlations of test scores obtained at different age levels on the same individuals in the course of longitudinal studies. Such curves can be interpreted as reflecting the rate of mental growth but say nothing as to the nature of what is developing. Indeed, if intelligence is defined in the narrow sense of the abilities to generalize and abstract, then any qualitative differences in these abilities will necessarily be obscured by the curve of mental growth which suggests merely a quantitative increase in mental ability with increasing age.

Looked at from the standpoint of Piagetian psychology, however, mental growth involves the formation of new mental structures and consequently the emergence of new mental abilities. The child, to illustrate, cannot deal with propositional logic of the following sort, "Helen is shorter than Alice and taller than Ethel, who is the tallest of the three?" (Glick & Wapner, 1968), nor can children grasp the metaphorical connotations of satirical cartoons or proverbs (Shaffer, 1930). Adolescents, in contrast, have no trouble with either propositional logic or with metaphor. In the Piagetian view, therefore, mental growth is not a quantitative but rather a qualitative affair and presupposes significant differences between the thinking of children and adolescents as well as between preschool and school age children.

These qualitative differences are, as a matter of fact, built into the items of mental tests but are masked by the assignment of point scores to successes and failures. On the Wechsler Intelligence Scale for Children various of the sub-tests recognize qualitatively different responses only by assigning them additional points (Wechsler, 1949). For example, a child who says that a peach and a plum are alike because "they both have pits" is given a single point, whereas a child who says "they are both fruit" is given two points. On other sub-tests, such as the arithmetic sub-test, there is no point differential for success on problems which patently require different levels of mental ability. To illustrate, correct answers to the following two problems are both given only a single point: "If I cut an apple in half, how many pieces will I have?" A correct answer to that question is given the same score as the correct answer to this problem:

Smith and Brown start a card game with $27 each. They agree that at the end of each deal the loser shall pay the winner one third of what he (the loser) then has in his possession. Smith wins the first three deals. How much does Brown have at the beginning of the fourth deal?

Clearly, the items on any given sub-test can tap quite different mental processes but these qualitative differences are obscured by assigning equivalent point scores to the various items regardless of the mental processes involved.

This is not to say that Piaget is right and that the mental test approach is wrong, or vice versa. The quantitative evaluation of mental growth is necessary and has considerable practical value in predicting school success. The qualitative approach is also of value, particularly when diagnosis of learning difficulties and educational remediation are in question. Which approach to mental growth one adopts will depend upon the purposes of the investigation. The only danger in the quantitative approach is to assume that, because sub-tests include items of the same general type and are scored with equal numerical weights, that they therefore assess only quantitative differences in the ability in question.

The Contributions of Nature and Nurture to Intelligence

Still a third way in which the psychometric and Piagetian views of intelligence differ has to do with the manner in which they treat the contributions of nature and nurture to intellectual ability. In the psychometric approach this contribution is treated substantively, with regard to the amount of variance in intellectual ability that can be attributed to nature and nurture respectively. Piaget, on the contrary, treats these contributions functionally with respect to the regulative role played by the environment or inner forces for any given mental activity. Both positions now need to be described in somewhat more detail.

The psychometric approach is substantive (and static) in the sense that it regards intelligence as capable of being measured and holds that such measures can be used to assess the extent to thich nature and nurture contribute to intellectual ability. In the discussion of genetic causality the various components into which test scores could be analyzed were briefly noted. We are indebted to writers such as Burt & Howard (1957) and Jensen for making clear the many and complex determinants into which test performance can be analyzed. Without wishing to minimize these other determinants, the needs of the present discussion will be served if we consider only how the psychometric approach arrives at the contribution of the heredity and environmental factors.

As Jensen points out, heritability is the proportion of variability among observed or phenotypic intelligence (test scores) that can be attributed to genotypic variations. Estimates of heritability are obtained from correlational data for subjects with known kinship relations such as parents and children, siblings, and identical twins. The contribution of the environment is arrived at somewhat differently. Variability in intelligence test scores attributable to the environment is estimated from that variability which cannot be attributed to any other factors. It is, in fact, the residual variance, that which is left after all the other factors contributing to intelligence test performance have been accounted for. For the psychometrician, then, nature and nurture are regarded as substantive and static, and their contributions are assessed quantitatively with the aid of statistical procedures.

When we turn to the work of Piaget, however, we encounter quite a different conception of the contributions of nature and nurture. In Piaget's view, these contributions must be conceived functionally and dynamically with respect to their regulatory control over various mental activities. In this regard Piaget's views are not unlike those of David Rapaport (1958) who spoke of "the relative autonomy of the ego," a conception which may help to introduce Piaget's somewhat more difficult formulation. Rapaport argued that we are endowed with some mental processes, such as perception, that are responsive to the environment and so tend to guarantee or insure a certain independence of the mind from the domination of instinctual drives. Other mental processes, such as fantasy, are most responsive to internal forces and these in turn guarantee a certain independence of the mind from the domination of the environment. The presence and activity of both types of processes thus insures that the mind is enslaved neither by the environment nor by drives but retains a "relative autonomy" from both.

Piaget's view (1967c) is roughly similar. He argues that intelligence is an extension of biological adaptation which, in lieu of the instinctive adaptations in animals, permits relatively autonomous adaptations which bear the stamp not only of our genetic endowment, but also of our physical and social experience. On the plane of intelligence we inherit the processes of assimilation (processes responsive to inner promptings) and of accommodation (processes responsive to environmental intrusions). Assimilative processes guarantee that intelligence will not be limited to passively copying reality, while accommodative processes insure that intelligence will not construct representations of reality which have no correspondence with the real world. To make this functional conception of the contributions of nature and nurture to intelligence concrete, let us consider several different mental abilities which are differently regulated by internal and external forces.

If we look at imitation (Piaget, 1951), it is clear that it is largely accommodative in the sense that it is most responsive to environmental influence and is relatively independent of inner forces. The vocal mimic, for example, is expert to the extent that he can capture the pitch, timbre and inflections of his model's voice and to the extent to which he can suppress those aspects of his own speech which differ from the model's. Play, in contrast, is largely assimilative in that it is most responsive to inner needs and is relatively independent of environmental influence. The child who uses a stick alternatively as a gun, as an airplane and as a boat has responded to the object largely in terms of his own inner needs and with a relative disregard of its real properties.

Between the two extremes of imitation and play is intelligence which manifests a balance or equilibrium between assimilative and accommodative activities and is thus relatively autonomous both of inner *and* outer forces. To illustrate, suppose we deduce, from the premise that Helen is taller than Jane and that Jane is taller than Mary, that Helen is the taller of three girls. We have in so doing attained a new bit of knowledge, an adaptation, but without altering the elements involved (assimilation without transformation of the objects) and without modifying the reasoning processes (accommodation without alteration of mental structures). Reason, or intelligence, is thus the only system of mental processes which guarantees that the mind and the environment will each retain its integrity in the course of their interaction.

Accordingly, for Piaget as for Rapaport, the question is not how much nature and nurture contribute to mental ability, but rather the *extent to which various mental processes are relatively autonomous from environmental and instinctual influence.* Such a conception is functional and dynamic, rather than substantive and static, because it deals with the regulatory activity of nature and nurture upon various mental processes. Those processes which show the greatest independence from environmental *and* internal regulation, the rational processes, are the most advanced of all human abilities. It is for this reason that Piaget reserves for them, and for them alone, the term intelligence.

In summary then, the psychometric and Piagetian approaches to intelligence differ with respect to: (a) the type of genetic causality which they presuppose; (b) their conceptions of the course of mental growth; and finally (c) the manner in which they conceive the contributions of nature and nurture to intellectual ability. In closing this section on the differences between the two positions I want to say again that the differences arise from differences in perspective and emphasis and are not contradictory but rather complementary. Both the psychometric and the Piagetian approaches to the conceptualization of human intelligence provide useful starting points for the assessment and interpretation of human mental abilities. Let us turn now to a couple of practical issues related to the modification and stimulation of mental abilities.

Practical Issues

In his essay, Jensen has tried to clarify many of the ambiguities regarding the nature and modification of intellectual ability and to put down some of the myths and misinterpretations prevalent with regard to test intelligence. For the most part, I find myself in agreement with Jensen and in this section, I would like to discuss two practical issues related to the modification and stimulation of intellectual abilities which seem to involve some misinterpretation of the Piagetian position. First, Piaget's insistence upon the qualitative differences between the modes of thinking at different age levels has been wrongly taken to suggest the need for preschool instruction in order to move children into concrete operational stage more quickly. Secondly, Piaget's emphasis upon the non-chance or self-directed nature of mental development has mistakenly been taken as justification for the use of methods such as "discovery learning" which supposedly stimulate the child's intrinsic motivations to learn. I would like, therefore, to try in the following section to clarify what seems to me to be the implications of Piaget's conception of intelligence for preschool instruction and for the implementation of intrinsic motivation.

Preschool Instruction

There appears to be increasing pressure these days in both the popular and professional literature for beginning academic instruction in early childhood, i.e., from 3 to 5 years. Bruner's famous statement that "We begin with the hypothesis that any subject can be taught effectively in some intellectually honest form to any child at any stage of development" (Bruner, 1962, p. 33) as well as the work of Hunt (1961), of Bloom (1964), of O. K. Moore (1961), of Fowler (1968), and of Skeels (1966) have all been used in the advocacy of preschool instruction. Indeed Piaget and Montessori have been invoked in this connection as well. The argument essentially is that the preschool period is critical for intellectual growth and that if we leave this period devoted to fun and games, we are lowering the individual's ultimate level of intellectual attainment. Parental anxiety and pressure in this regard have been so aroused that legislation has been passed or is pending for the provision of free preschool education for all parents who wish it for their children in states such as New York, Massachusetts and California.

What is the evidence that preschool instruction has lasting effects upon mental growth and development? The answer is, in brief, that there is none. To prove the point one needs longitudinal data on adults who did not have preschool instruction but who were equal in every other regard to children receiving such instruction. With the exception of the Montessori schools, however, the preschool instruction programs have not been in existence long enough to provide any evidence on the lastingness of their effects. Indeed, most of the earlier work on the effects of nursery school education (see Goodenough, 1940, and Jones, 1954, for reviews of this literature) has shown that significant positive effects are hard to demonstrate when adequate experimental controls are employed. It is interesting that no one, to my knowledge, has done a longitudinal study of adult Montessori graduates. Have they done better in life than children from comparable backgrounds not so trained? In any case, it is such unavailable longitudinal data that is crucial to the proposition that the preschool period is a critical one for intellectual development.

I am sure that someone will object at this point that studies of mental growth such as those of Bloom (1964) suggest that half of the individual's intellectual potential is realized by age four. Does this not mean that the preschool period is important for intellectual growth and that interventions during this period will have lasting effects? Not necessarily, if we look at the facts in a somewhat different way. Bloom writes, "Both types of data suggest that in terms of intelligence measured at age 17, about 50% of the development takes place between conception and age 4, about 30% between ages 4 and 8, and about 20% between 8 and seventeen" (Bloom, 1964, p. 88). Now an equally feasible implication of this statement is quite in contradiction to that of preschool instruction: the child has only 50% of his intellectual ability at age 4 but 80% at age 8, why not delay his education three years so that he can more fully profit from instruction? With 80% of his ability he is likely to learn more quickly and efficiently and is not as likely to learn in ways that he will need to unlearn later. That is to say, without stretching the fact, it is possible to interpret the Bloom statement as implying that instruction should *not* be introduced into the preschool program.

Not only is there no clear-cut longitudinal data to support the claims of the lastingness of preschool instruction, there is evidence in the opposite direction. The work cited by Jones (1954) and by Piaget (1967b) in the quotations given earlier in this paper are cases in point. This evidence, together with more recent data reported in Jensen's paper, suggest a negative correlation between early physical maturation and later intellectual attainments. Animals are capable of achieving early some skills (a dog or a chimp will be housebroken before a child is toilet trained) but perhaps at the expense of not being able to attain other skills at all. This data suggests the hypothesis that *the longer we delay formal instruction, up to certain limits, the greater the period of plasticity and the higher the ultimate level of achievement.* There is at least as much evidence and theory in support of this hypothesis as there is in favor of the early-instruction proposition. Certainly, from the Piagetian perspective, there are "optimal periods" for the growth of particular mental structures which cannot be rushed.

Please understand, I am not arguing against the benefits of preschool enrichment for children. Even preschool instruction may be of value for those disadvantaged children who do not benefit from what Strodtbeck (1967) called the "hidden curriculum of the middle class home." What I am arguing is that there is no evidence for the *long term effects* of either preschool instruction or enrichment. Nursery school experience most assuredly has immediate value for the child to the extent that it helps him to appreciate and enjoy his immediate world to the full and to better prepare him for future social and intellectual activities. Everyone, for example, recognizes the value of a vacation without expecting that it will produce any permanent alterations. Isn't it enough that we lighten the burdens of childhood for even a brief period each day without demanding at the same time that we produce permanent results? The contributions of the nursery school, no less than that of the vacation, do not have to be long-lived to be of value.

In closing the discussion, I would like to emphasize another side to this issue of preschool instruction. This is the consideration that the emphasis on preschool education has obscured the fact that it is the elementary school years which are crucial to later academic achievement. It is during these years that the child learns the basic tool subjects, acquires his conception of himself as a student and develops his attitudes towards formal education. In this connection it might be well to quote a less publicized finding of Bloom's (1964) study:

We may conclude from our results on general achievement, reading comprehension and vocabulary development, that by age 9 (grade 3) at least 50% of the general achievement pattern at age 18 (grade 12) has been developed whereas at least 75% of the pattern has been developed by age 13 (grade 7). (Bloom, 1964, p. 105)

With respect to the intellectual operations of concern to Piaget, similar trends appear to hold true. While children all over the world and across wide ranges of cultural and socioeconomic conditions appear to attain concrete operations at about the age of 6 or 7 (Goodnow, 1969), the attainment and use of formal operations in adolescence, in contrast, appear to be much more subject to socioculturally determined factors such as sex roles and symbolic proficiency (Elkind, 1961; Elkind, Barocas & Rosenthal, 1968; Goodnow & Bethon, 1966). Apparently, therefore, environmental variation during the elementary school period is more significant for later intellectual attainments of the Piagetian variety. In short, there is not much justification for making the preschool the scapegoat for our failures in elementary education. Like it or not, the years from six to twelve are still the crucial ones with respect to later academic achievement.

Motivation and Intellectual Growth

In recent years there has been an increasing recognition among psychologists such as Berlyne (1965), Hunt (1965), and White (1959), that certain mental activities can be self-rewarding and do not have to be externally reinforced. European writers such as Piaget (1954) and Montessori (1964) long ago recognized the existence of "intrinsic motivation" (to use Hunt's apt phrase), and Montessori in particular gave incomparable descriptions of children who suddenly discover they can read and proceed to read everything in sight. Piaget (1967d) too, has argued that needs and interests are simply another aspect of all cognitive activities.

Educators, however, in their efforts to capitalize upon this intrinsic motivation seem to have missed the point of what Montessori and Piaget had in mind. To maximize intrinsic motivation and to accelerate mental growth we have recently had an emphasis upon "learning by discovery" and upon "interesting reading materials" and so on. These approaches miss the point because they assume that intrinsic motivation can be built into materials and procedures which will in turn maximize mental growth. But as Piaget and Montessori pointed out (Elkind, 1967) intrinsic motivation resides in the child and not in methods and procedures. It is the child who must, at any given point in time, choose the method of learning and the materials that are reinforcing *to him*. Without the opportunity for student choice and the provision of large blocks of time in which the child can totally engross himself in an activity, the values of intrinsic motivation will not be realized.

Indeed, I am very much afraid that by the time most children have reached the third or fourth grade a good deal of their intrinsic motivation for learning has been stifled. This is because spontaneous interest follows only the timetable of the child's own growth schedule. We can all remember, I am sure, those periods when we were so totally immersed in an activity that we forgot time, food and rest. During such periods we are at our creative and productive best and afterwards the feeling of exhaustion is coupled with a deep sense of accomplishment. In the school, however, we do not permit children to become totally engrossed in an activity but rather shuttle them from activity to activity on the hour or half hour. The result is what might be called *intellectually burned children*. Just as the burned child shuns the fire so the intellectually burned child shies away from total intellectual involvement.

How is this condition produced? In clinical practice we often see children (and adults) who are unwilling to form any emotional attachment. In the history of such children one always finds a series of broken relationships due to a wide variety of causes including the death of parents or the forced separation from them. Such children have learned that every time they reached out and became emotionally involved, rejection, hurt and misery were the result. Consequently they prefer not to get involved any more because the pain and anguish of still another broken relationship is just too high a price to pay for an emotional attachment. The intellectually burned child is in somewhat the same position. He refuses to become totally involved in intellectual activities because the repeated frustration of being interrupted in the middle is just too much to bear. Our lockstep curricula, thirty minutes for this and an hour for that, have the consequence, I suspect, of producing children who shun the fire of intense mental involvement.

Accordingly, the educational practice which would best foster intrinsically motivated children in the Piagetian and Montessori sense would be the provision of "interest areas" where children could go on their own and for long periods of time. Only when the child can choose an activity and persist at it until he is satiated can we speak of true intrinsically motivated behavior. Where such interest areas and time provisions have been made, as in the World of Inquiry School in Rochester, New York, the results are impressive indeed.[1]

In summary then, the Piagetian conception of intelligence provides no support either for those who advocate formal preschool instruction or for those who argue for new methods and materials to stimulate intrinsic motivation. As we have seen, there is no evidence as yet for the lastingness of preschool instruction. In addition, intrinsic motivation seems best stimulated by allowing the child to engage in the activity of his choice for unbroken periods of time. As Jensen has so rightly pointed out, if we really want to maximize the effects of instruction, it does not pay to blink at the facts whether they have to do with racial or socioeconomic differences in intelligence, the effects of preschool instruction, or the nature of intrinsic motivation.

[1] The results of our preliminary evaluation of this school suggest that World of Inquiry pupils are significantly higher in their need for achievement and more positive in their self evaluations than are their matched controls (children taken from the waiting list) who are attending other schools.

References

Berlyne, D. E. Curiosity and education. In J. D. Krumboltz (Ed.) *Learning and the educational process.* Chicago: Rand McNally, 1965, 67-89.

Bertalaffny, Ludwig von. *Modern theories of development.* New York: Harper & Bros. (Torchbook Ed.) 1962.

Bloom, B. S. *Stability and change in human characteristics.* New York: John Wiley & Sons, Inc., 1964.

Bruner, J. *The process of education.* Cambridge, Mass.: Harvard University Press, 1962.

Burt, C. *Mental and scholastic tests.* London: Staples Press, 1962 (4th edition).

Burt, C., & Howard, M. The relative influence of heredity and environment on assessments of intelligence. *British Journal of Statistics Psychology,* 1957, **10**, 33-63.

Elkind, D. Quantity conceptions in junior and senior high school students. *Child Development,* 1961, **32**, 551-560.

Elkind, D. Piaget and Montessori. *Harvard Educational Review,* 1967 (Fall) 535-545.

Elkind, D., Barocas, R., & Rosenthal, B. Combinatorial thinking in children from graded and ungraded classrooms. *Perceptual and Motor Skills,* 1968, **27**, 1015-1018.

Fowler, W. The effect of early stimulation in the emergence of cognitive processes. In R. D. Hess & Roberta M. Meyers (Eds.) *Early Education.* Chicago: Aldine Press, 1968, 9-36.

Glick, J., & Wapner, S. Development of transitivity: Some findings and problems of analysis. *Child Development,* 1968, **39**, 621-638.

Goodenough, Florence. New evidence on environmental influence on intelligence. *Yearbook of the National Society for the Study of Education,* 1940, **39**, 307-365.

Goodnow, Jacqueline J. Problems in research on culture and thought. In D. Elkind and J. Piaget (Eds.) *Studies in cognitive development.* New York: Oxford University Press, 1969, 439-464.

Goodnow, Jacqueline J., & Bethon, G. Piaget's tasks: The effects of schooling and intelligence. *Child Development,* 1966, **37**, 573-582.

Hunt, J. McV. *Intelligence and experience.* New York: The Ronald Press, 1961.

Hunt, J. McV. Intrinsic motivation and its role in psychological development. In D. Levine (Ed.) *Nebraska symposium on motivation.* Lincoln: University of Nebraska Press, 1965, 189-282.

Jones, H. E. The environment and mental development. In L. Carmichael (Ed.) *Manual of child psychology.* New York: John Wiley & Sons, Inc., 1954, 631-696.

Kittlewell, H. B. D. Selection experiments on industrial melanism in the lepidoptera. *Heredity,* 1955, **9**, 323-342.

Montessori, Maria. *The Montessori Method.* New York: Schocken, 1964 (first published in English, 1912).

Moore, O. K. Orthographic symbols and the preschool child: A new approach. In E. P. Torrence (Ed.) *Creativity: 1960 proceedings of the 3rd conference on gifted children.* Minneapolis: University of Minnesota, Center for Continuation Studies, 1961.

Piaget, J. *Play, dreams and imitation in childhood.* New York: Norton, 1951.

Piaget, J. *Les relations entre l'affectivité et l'intelligence dans la developpement mental de l'enfant.* Paris: C.D.U., 1954 (mimeographed and bound lectures given at the Sorbonne).

Piaget, J. Genesis and structure in the psychology of intelligence. In D. Elkind (Ed.) *Six Psychological Studies by Jean Piaget.* New York: Random House, 1967a, 143-158.

Piaget, J. *On the nature and nurture of intelligence.* Address delivered at New York University, March, 1967b.

Piaget, J. Intelligence et adaptation biologique. In F. Bresson *et al* (Eds.) *Les Processus d'adaptation,* Paris: Presses Universitaires de France 1967c, 65-82.

Piaget, J. The mental development of the child. In D. Elkind (Ed.) *Six Psychological Studies by Jean Piaget.* New York: Random House, 1967d, pp. 3-73.

Rapaport, D. The theory of ego autonomy. *Bulletin of the Menninger Clinic,* 1958, **22,** 13-35.

Shaffer, L. F. Children's interpretations of cartoons. *Contributions to Education,* No. 429. New York: Teacher's College, Columbia University, 1930.

Skeels, Harold M. Adult status of children with contrasting early life experiences. *Monographs of the Society for Research in Child Development,* 1966, **31,** 3, No. 105.

Spearman, C. *The nature of "intelligence" and the principles of cognition.* London: Macmillan, 1923.

Strodtbeck, F. L. The hidden curriculum of the middle class home. In H. Passow, Miriam Goldberg and E. J. Tannenbaum (Eds.) *Education of the disadvantaged.* New York: Holt, Rinehart & Winston, 1967, 244-259.

Waddington, C. H. *The nature of life.* New York: Atheneum, 1962a.

Waddington, C. H. *How animals develop.* New York: Harper & Bros. (Torchbook Ed.), 1962b.

Wechsler, D. *The measurement of adult intelligence.* Baltimore: Williams & Wilkens, 1944.

Wechsler, D. *Wechsler intelligence scale for children.* New York: Psychological Corporation, 1949.

White, R. W. Motivation reconsidered: The concept of competence. *Psychological Review,* 1959, **66,** 297-333.

Heredity, Environment, and Educational Policy

LEE J. CRONBACH, *Stanford University*

Professor Cronbach accepts some but by no means all of Professor Jensen's empirical conclusions. In the following review he indicates some research that bears on their points of disagreement. Cronbach suggests that such distinctions as Jensen's dichotomy between "Level I" and "Level II" abilities over-simplify the many dimensions of individual differences, and he disagrees with the educational policy he feels is implied by Jensen's recommendations for education. Beyond this, Professor Cronbach poses a more basic question—"Intelligence for what?"—a question of the compatibility of current social aims of schooling with long range changes in our social and technological structure.

Professor Jensen is among the most capable of today's educational psychologists. His research is energetic and imaginative. In the present paper, an impressive example of his thoroughness, I am sure every reader has had my experience of encountering valuable information in areas where he thought himself *au courant*. Unfortunately, Dr. Jensen has girded himself for a holy war against "environmentalists," and his zeal leads him into over-statements and misstatements. Rather than list the points where Dr. Jensen and I agree, and those where we diverge, let me begin with an integrated statement of my view on the major themes.

I do not doubt that performance—intellectual, physical, or social—is developed from a genotypic, inherited base. The organism, as it evolves prenatally and post-

* This comment was prepared under support from the U. S. Office of Education, but the views are those of the writer only.

Harvard Educational Review Vol. 39 No. 2 Spring 1969, 338–347

natally, incorporates energy and information. What the person does with an experience, and what it does to him, depends on physical structures that were laid down during the previous years, or days, of his existence. Human development is a cumulative, active process of utilizing environmental inputs, not an unfolding of genetically given structures (Caspari, 1968).

The genetic populations we call races no doubt have different distributions of whatever genes influence psychological processes. We are in no position to guess, however, which pools are "inferior." Such a comparison is not meaningful, except in terms of the probability that the member of the group will be able to cope in some specific way with some specific challenge, after he has developed for a specified period in some specified environment.

Darwin's catchy phrase, "survival of the fittest," has misled hereditarians for a century. A genetic factor that has survival value in one environment is detrimental in another. Whatever the individual's genotype—barring gross defects—there are environments in which he will develop so as to function well and others in which he will develop poorly. For another genotype these effects may be reversed. There are many possible educational and developmental environments. At any age, the person is the phenotypic product of his genotype and his experiences to date; this phenotype may make him unready to profit adequately from the treatments now established for persons at that age. He might, however, be well equipped for some other series of educational procedures we could devise.

The phrase "improve the environment," born of the enthusiasm of the Social Darwinists, has misled environmentalists for two generations. Environments cannot be arrayed from good to bad, rich to poor. The highly stimulating environment that most of us think of as "rich" promotes optimal growth for some persons and may not be suitable for others. Environments can be varied along many dimensions, and the optimum with respect to each dimension depends on the person's phenotype at a given time. We think of the infant as deprived when he has nothing to gaze upon but a blank ceiling, but nothing is gained by making the environment so richly patterned that he cannot direct his attention. The pattern that holds attention varies with his age and his past experience (Fantz, 1961). An information-laden environment is rich, in some sense; but the right amount of redundancy and of detail depends on the learner's maturity. Conditions that make more information available may create an overload and so impair learning (Wicklegren and Cohen, 1962). How much stimulation is optimal, how much assistance, how much external monitoring and reinforcement, how much pressure for excellence, how much of the conceptual, and how much of the concrete—

these depend on the state of the individual and indirectly on his genotype. The optimum might be genetically determined; one can imagine, for example, metabolic differences that would make some children more impulsive than others and hence in need of a more calming environment. But the present state embraces biological structures, habits, attitudes, and meaning systems that are the residue of a long series of transactions. Some of them are transactions of genetically-determined structures with the environment, but more are transactions of the phenotype at a given moment with the environment of that moment.

There *has* been too much blithe optimism about our ability to improve the intellectual functioning of the slum child and the retarded child. Programs of compensatory education seem to have had no reliable and lasting effect. It may have been a sound political decision to launch massive compensatory programs, if only as a token of public concern. But far more was promised than we know how to deliver, and the hectic effort has drawn energies away from the needed basic, objective research. We need to clarify the aims of the programs and the hypotheses on which the experimental programs operate; this will move us beyond argument as to whether disadvantaged children can be helped, toward tested, workable plans (Hess & Bear, 1968).

There are two ways in which altered environment can be helpful, and these have been confused. One is to provide an optimal *maintenance* environment. That is how we promote the best growth of a plant; we select its planting spot, fertilizer, and so on to fit its requirements and maintain them throughout its life. This is likewise our way of overcoming the deficiency of the PKU child; we modify the environment by means of a special maintenance diet. The other is to provide a special environment for a brief *intervention* period, in the hope that the person will be brought to normal readiness for the conventional environment so that the special treatment can be discontinued. Remedial intervention can supply needed skills, alter habits, and overcome many physical defects. Which of the two lines of attack is suited for the problem of inaptitude for the existing educational environment is in part an empirical question.

But much of the decision rests on little-recognized policy issues. The intervention treatment intends that pupils placed in it shall ultimately complete the regular program of socialization and be indistinguishable as a group from those who did not need special treatment. I am sympathetic with the objection of Gordon and Wilkerson, quoted by Jensen, that it is wrong-headed to try to make the slum child fit the middle-class stereotype, as child or adult. But education must have a clear idea of its intended product. If we are to bring these children to a self-respecting adulthood, we must define for them a prospective role that has at least as great a value, to the individual and to society, as the middle-class model of industry, articulateness, social and cultural concern, and self-regulation. No one protesting against middle-classness has gone on to describe a possible, viable society in which large subsegments of society have radically different orientations and functions.

Today's discontent is a clamorous crisis that distracts us from a quieter, yet more ominous crisis—the bankruptcy of long-range social planning. Lacking visions of what society might become, we are training people for a *status quo* that is already vanishing. The schools are committed to training people for production, responsibility, creation, and leadership. The intervention programs seek to offer that way of life to all. But the fact is that automation, centralization, complexity, and abundance already have created a society where most people work less and less, while the manager and the professional work 50 to 70 hours per week. Huxley's beehive *World*, where a few highly educated persons, conditioned to self-denial, carry the productive burden, is already the American way of life. It is against that world—where the uptight Alphas are the slaves—that our brightest youth are protesting. The time has come for far less concern with the total man-years of education produced by our system, and for intensive and sober concern with the capital question, "Intelligence for what?"

It is hard to see how evidence on heritability provides a base for social policy. It is surely humane policy—without regard to questions of heritability—to facilitate birth control. It is inconceivable that we will scale welfare payments to penalize the child in a large family. I hope it is inconceivable that data on heredity—whether of the individual or the group—will persuade us that some children should be taught concepts, some taught rote verbal associations, and some taught how to change tires. Jensen seems to argue that the disadvantaged should be taught by rote methods. But the cut-and-dried answers that can be learned by rote are not the answers that one needs if he is to cope with a changing world and to live an appreciative and expressive life. The proper and necessary strategy is to find alternative means of bringing all children as far as we can toward self-fulfillment. Under our present conception of the good life we cannot set goals of entirely different character for different pupils. It is regrettable that Jensen says little about the policies he would expect us to follow if we accepted his empirical conclusions.

On the scientific side, it is vital to break away from such stereotyped terms as "intelligence" and "learning ability." There is a spectrum of performances, ranging from crystallized, overlearned routines to fluid information-processing, often referred to as g (Cronbach, 1969). Fluid ability is measured in tests like the maze, the matrix or figure analogies, block design, and embedded figures; Jensen's "Level II" abilities involve it. Crystallized abilities are diverse and specific: spelling of -gn words, handling of subjunctive clauses, etc. In schools as they now are, success is best predicted by taking inventory of the relevant crystallized abilities with which the pupil starts the year. The verbal "intelligence" test succeeds as a predictor primarily because it reflects concrete achievements. A child with average fluid ability and low crystallized ability is likely to do poorly; we have never succeeded in devising a mass educational program in which such a child is likely to achieve average success. Analytic ability should be a resource on which education builds, and as of now it is not.

Because learning abilities are plural, they are not adequately conceptualized by Jensen's Level I-Level II system. Many processes contribute to effective learning; some are under conscious control or trainable, and some not. Which processes are required depends on what is being learned and what kind of instruction is employed. At times, striking differences in "ability" can be overcome very simply. Lower-class children are inferior in paired-associate learning, according to many studies—no doubt the task would have a fairly substantial heritability (H) index.[1] But simply coaching the lower-class children to make up "meaningful" associations for the word pairs brought them up to the middle-class rate of learning. This finding, coming from Jensen and Rohwer (1965), seems to contradict what Jensen says in this paper. It is at least possible that on the Glasman "Level II" task (recall of objects that can be conceptually clustered) the lower-class children could overtake the middles if made aware of the usefulness of analysis. Indeed—stop the presses!—a brand-new study seems to demonstrate cleanly that very simple instruction does overcome initial weakness on the Glasman task (Moely *et al.*, 1969). Capability is not at issue when a child does not call upon an ability he possesses.

As to heritability, there is less here than meets the eye. The term, though standard in genetics, is mischievous in public discussion, for it suggests to the unwary that it describes the limit to which environmental change *can be* influential. Not so. The H index describes a certain population, having a certain gene pool and having developed in a certain range of environments. (There are some

---

[1] Heritability is "a population statistic, describing the relative magnitude of the genetic component (or set of genetic components) in the population variance of the characteristic in question." (Jensen, p. 42). Cf. Jensen pp. 42-43 for a formal definition of the term—ed.

treacherous assumptions, well explained by Huntley, 1966; the most critical of these is that environments are distributed at random over the various genotypes. But even if one made alternative assumptions, the H value would surely remain above 0.50, and it is hard to see how moderate changes in the index would alter one's social policy.) A phenotype that is 100 percent heritable is not affected by the variations among existing environments. But introduce a "mutant" environment, and H will change; this is exactly what happened in the case of PKU—a direct genotype-phenotype link was broken. Likewise, note the report of Osborne and Gregor (1966) that a certain type of spatial test has an H value of 0.89, alongside the finding (Brinkmann, 1966) that as soon as someone made an effort to train for this kind of ability, scores were increased by large amounts. The influence of environment on a trait with high H is also dramatically apparent to the American Fulbrighter of average height who finds large numbers of today's Japanese youth towering over him. Pool the heights of 1940 Japanese and 1970 Japanese in a single calculation, and H will be quite a bit lower than it is in either group alone. In most cultures mental-test scores show similar generation-to-generation gains attributable to environment.

Attention should be directed to Jensen's remark that environment does not affect stature above the level that includes "minimal daily requirements of minerals, vitamins, and proteins" (p. 60). But that is the point. Nutritional science now tells us to include certain chemical substances in the diet; these give heredity something to work with. You do not increase stature two inches in a generation just by providing more and more rice. You do not increase mental ability just by providing more stimulation. Analytic research will in due time specify the needed ingredients in an educational diet.

The heritability index of 0.80 is impressive, but it is less discouraging than Jensen implies; environments of the sort we now have can improve ability, if we can choose the environment to fit the individual instead of relying on fortuitous correspondences. A brief technical sortie will put a new light on the index of 0.80. Think of an "expected IQ"—the hypothetical average IQ of a thousand persons having identical genes, who have been assigned at random to environments. Assume that all IQs are "true" scores, perfect measures. Starting with Jensen's H value, the correlation of individual IQ with expected IQ, over the population, is $(0.80)^{1/2}$. The standard error of estimate for individual IQ is approximately $[200(1-0.80)]^{1/2}$ or 6.3. Hence persons having the same genes are distributed over an IQ range of more than 25 points. With run-of-the-statistics cases and within the range of present environments, the individual who draws an environment fitted to his genotype develops an IQ some 6 points better than the expected IQ for that genotype, and 12 or more points better than does one who is unlucky in the draw. If an effect of this size could be brought under control and applied population-wide, it would surely be economically and culturally beneficial.

It is necessary to deal summarily with a number of aspects of the paper. I have detected substantial distortions in Jensen's report of some research, and I must therefore warn the reader against accepting his summaries. Selective breeding studies are a case in point; Jensen says that "maze learning ability" can be bred (p. 30). But Anastasi, interpreting the same data, emphasizes that the superiority of the selected stock was *not* due to any superior "learning ability" (1958, p. 91). In fact, some of the studies were carried out precisely to demonstrate that breeding selects on particular temperamental traits that facilitate learning under one condition and impede it under others. The maze learning superiority of the Tryon strains was specific to one kind of maze under one kind of incentive. I particularly invite the reader's attention to John Paul Scott's eloquent attack on the idea of a general inherited learning ability (see Rosenblith and Allinsmith, 1966, pp. 54-57), since Jensen cites Scott as if Scott endorsed such an idea (p. 30).

Jensen is severe with studies that encourage the belief that the retarded and the disadvantaged can be helped. He is right to be critical of many of the studies that claim positive results. He could well have cited the Zigler-Butterfield (1968) demonstration that simple increases in motivation for the test account adequately for most reported before-and-after differences in preschool children. He could justly have been more severe in disposing of the Rosenthal-Jacobsen study—which purports to find evidence that giving the teacher mental-test data biases the teacher's handling of the pupil. He gives excellent advice (pp. 96 ff.) on the design of evaluative studies. So far, so good. But when he cites the Wheeler study of a gain in IQ following the opening of the Tennessee hills to the modern world, around 1930, he goes out of his way to say, "The decline in IQ from age 6 to age 16 was about the same in 1940 (from 103 to 80) as in 1930 (from 95 to 74)." More accurately, the 16-year-olds declined from 95 at age 6 in 1930 to 80 in 1940. These adolescents were dropping behind the norm group, most likely because their schooling was not up to that of the norm group. Jensen notes (p. 17) that Bloom summarizes age-to-age correlations of mental-test scores. It seems to me that, having introduced this source, Jensen was obligated to disclose that Bloom gives these data an interpretation opposite to Jensen's. Bloom sees the gains from year to year in test scores as random and unpredictable, hence due to external events and not inheritance. (This is one of several alternative interpretations that fit the data.)

There is plentiful evidence that late blooming occurs, i.e., that some persons rise dramatically in their relative position even as late as adolescence. To label these important effects as a "regression" effect (p. 99) neither explains nor diminishes them. There is a trivial regression effect, arising from sheer error of measurement on the earlier test; but most reports on large IQ changes indicate that the relatively low initial status was confirmed by several tests. Whenever prediction is imperfect because something has really happened between pre-test and post-test, there is some tendency for regression toward the mean, but that is no more than a paraphrase of the obvious: likely events happen more often than rare events.

Jensen accuses writers on education of underplaying or denying the role of heredity. Some of this bias does exist, but Jensen is unfair. He does not quote the writers in psychology and education who do devote space to heredity. And he does not see that, in writings for educators, it is pointless to stress heredity. The educator's job is to work on the environment; teaching him about heredity can do no more than warn him not to expect easy victories. Heritability of individual differences is not his concern. Even if, after education, rankings in ability were to correlate *perfectly* with some measure on the pupil's ancestors, the educator ought to be providing the best possible instruction he can for every pupil he faces. To be sure, the educator who makes policy has to decide, in allocating resources, whether to put more resources on the laggards or on the leaders, but this decision has to be based on a judgment about utilities. The same considerations enter this judgment, whether we assume zero heritability or perfect heritability.

Let me be telegraphic in disposing of some further reservations. Jensen states that "while fluid intelligence attains its maximum level in the late teens and may even begin to decline gradually shortly thereafter, crystallized intelligence continues to increase gradually with the individual's learning and experience all the way up to old age" (p. 13). I do not believe there is adequate evidence to offer a conclusion as to the trend of fluid ability with age. On another point, Jensen protests that we should not "reify g [general intelligence] as an entity" (p. 9), but it seems to me that he does so, especially as he begins to insist that it is a "biological reality" (p. 19). Fluid ability is demonstrated through a complex set of acts: attending, analyzing, encoding, transforming, etc. The process is not unitary even though the processes tend to be acquired through the same activities and, so, to be correlated. Later Jensen concurs in Zigler's criticism of "unbridled environmentalists" in whose writing "the concept of capacity is treated as a dirty word" (p. 29). But "capacity" *is* a dirty word, incapable of being given meaning and overwhelmingly capable of confusing discussions. It and all words like it refer to nothing but an expectancy under present circumstances. Intellectual capacity is continually being expanded by technological devices. Perhaps Zigler and Jensen will protest that the computer has not really increased man's mathematical capacity (though he now can solve in a day problems that once took a lifetime). Do they not admit that the long-ago invention of a spoken language increased "capacity"? And if so, where can they draw a line?

Jensen does not present clearly the important concepts of covariance and interaction (pp. 38 ff.). Covariance exists whenever persons of a certain genotype experience anything other than a random selection of the environments; nothing about matching "good" environment to "good" heredity is implied. Interaction exists when a difference in treatments produces one difference in outcomes in persons of one genotype, and some other difference in outcomes with a second genotype. Jensen offers as an example of interaction the possibility that genetically different individuals will gain different amounts of weight when given the same number of calories. "Their constitutions cause them to metabolize the same intake differently" (p. 40). This is a poor example. One might paraphrase to say that Jensen thinks children who inherit good g "metabolize exactly the same environmental intake quite differently"—but his calculations take *that* as a main effect of heredity.

Finally, Jensen denies that there is severe deprivation in the home of the slum child (p. 61). I am no authority in these matters, but I have heard descriptions—e.g., of small girls locked into an apartment to keep them from the dangers of the street—that seem to qualify as severe deprivation. Bronfenbrenner (1967) asserts that the presence of severe deprivation in Negro homes is "an unwelcome but nonetheless inexorable reality." In addition to "the indifference and hostility of the white community," he believes that the child-rearing practices of American Negroes are "stubborn obstacles to achieving quality and equality in education" (p. 910). Jensen himself defines brilliantly a large part of what a child must learn before he is ready to participate effectively in presentday schooling (p. 7); the Negro child is often not given that training (Hess & Shipman, 1965).

It will be apparent that Dr. Jensen and I agree on many fundamentals. With regard to policy, we both believe that every intervention program has to stand on its demonstrated merits. I would not ask that it "raise the IQ," but I would ask that it raise readiness for schooling or promote intrinsically valuable achievement. We both urge that new kinds of instruction be devised to fit diverse patterns of ability. One goal of instruction, in my opinion, should be to develop fluid ability and conceptual learning ability. The undoubted significance of heredity must not deter researchers from trying to design procedures that will do this. Impossible things are happening every day.

References

Anastasi, A. *Differential psychology* (3rd edition). New York: Macmillan, 1958.

Brinkman, E. H. Programmed instruction as a means of improving spatial visualization. *Journal of Applied Psychology,* 1966, **50**, 179-184.

Bronfenbrenner, U. The psychological costs of quality and equality in education. *Child Development,* 1967, **38**, 909-925.

Caspari, Ernst. Genetic endowment and environment in the determination of human behavior: Biological viewpoint. *American Educational Research Journal,* 1968, **5**, 43-56.

Cronbach, Lee J. *Essentials of psychological testing* (3rd edition). New York: Harper & Row, 1969. In press.

Fantz, R. L. The origin of form perception. *Scientific American,* 204:5, May, 1961, 66-72.

Hess, Robert D. & Bear, Roberta M. (Eds.). *Early education.* Chicago: Aldine, 1968.

Hess, Robert D. & Shipman, Virginia C. Early experience and the socialization of cognitive modes in children. *Child Development,* 1965, **36**, 869-886.

Huntley, R. M. C. Heritability of intelligence. In J. E. Meade & A. S. Parker (Eds.). *Genetic and environmental factors in human ability.* Edinburgh: Oliver & Boyd, 1966, pp. 201-218.

Jensen, Arthur M. & Rohwer, W. D., Jr. Syntactical mediation of serial and paired-associate learning as a function of age. *Child Development,* 1965, **36**, 601-608.

Moely, Barbara E., Olson, Frances A., Halwes, Terry G., & Flavell, J. H. Production deficiency in young children's clustered recall. *Developmental Psychology,* **1**, 1969, 35-39.

Osborne, S. T., & Gregor, A. J. The heritability of visualization, perceptual speed, and spatial orientation. *Perceptual and Motor Skills,* 1966, **23**, 379-390.

Rosenblith, Judy F., & Allinsmith, Wesley. *The causes of behavior* (2nd edition). Boston: Allyn & Bacon, 1966.

Wicklegren, W. & Cohen, D. H. An artificial language and memory approach to concept attainment. *Psychological Reports,* 1962, **11**, 815-827.

Zigler, E. & Butterfield, Earl C. Motivational aspects of changes in IQ test performance of culturally deprived nursery school children. *Child Development,* 1968, **39**, 1-14.

A Letter from the South

WILLIAM F. BRAZZIEL, *Virginia State College*

This letter came, in its earliest form, even before subscribers had received their copies of the Winter issue. Mr. Brazziel had been following the Jensen controversy ever since Dr. Jensen spoke to the American Educational Research Association in 1968. When news of the Harvard Educational Review *article reached Mr. Brazziel through local publicity (newspapers and the coverage in* U. S. News and World Report) *the correspondence printed below began. In subsequent issues of the* Review *we expect to print further letters and comments from our readers.*

Sirs:

Thirteen years ago plaintiffs brought suit in Federal District Court to integrate the Louisiana public schools. The main argument of the defense attorneys and the superintendent of public instruction was that "white teachers could not understand the Nigra mind" and, therefore, would not be able to teach them effectively in integrated classrooms. The defense quoted heavily from the theories of white intellectual supremacy as expounded by Henry Garrett and Audrey Shuey

Last week, a scant five days after Arthur Jensen made headlines in Virginia papers regarding inferiority of black people as measured by IQ tests, defense attorneys and their expert witnesses fought a suit in Federal District Court to integrate Greensville and Caroline County schools. Their main argument was that "white teachers could not understand the Nigra mind" and that the Nigra children should be admitted to the white schools on the basis of standardized tests. Those who failed to make a certain score would be assigned to all black remedial schools where "teachers who understood them could work with them." The defense in this case quoted heavily from the theories of white intellectual supremacy as expounded by Arthur Jensen.

Harvard Educational Review Vol. 39 No. 2 Spring 1969, 348–356

It will help not one bit for Jensen or the HER editorial board to protest that they did not intend for Jensen's article to be used in this way. For in addition to superiority in performing conceptual cluster tricks on test sheets, the hard line segregationist is also vastly superior in his ability to bury qualifying phrases and demurrers and in his ability to distort and slant facts and batter his undereducated clientele into a complete state of hysteria where race is concerned.

Jensen and the HER editorial board will modestly admit that they have superior intellects and I am sure they realized the consequences of their actions. Questions now arise as to why they decided to raise this issue, in this way, and at this time.

Fortunately, doubts about the ability of black and yellow people to master war, finance, science and technology are waning rapidly in both white and black minds. The imprecision of standardized testing is now clear to most literate people and the criminal use to which they are put in schools is also becoming clearer. Black history has made people aware that white people did *not* give America such things as the stoplight, the shoe last, heart operations and sugar refining but that black people did this. That John Smith did not develop corn and tobacco but learned to grow these crops from the Indians. And the beat goes on. People are now witnessing with their very eyes the fact of black youth finally given a half of a chance at education and jobs and being able to make exotic formulas for bombs and napalm as well as anyone else. As a result of all of this, I think the present set-to might be the last go-round for white supremacy psychological theory.

I would hope the Jensenites could alter their stance and approach and try to bring some good out of this situation after all. They might work their way out of ethnic learning styles by broadening their research to include all ethnic groups. We have some rather learned men in our area who believe that English-Americans are atop the pyramid of abstract learning abilities with Welsh, German, French, Belgian, Norwegian, Swiss, Finnish, Danish and Swedish occupying the next nine rungs in the order listed. After the top ten have been given their just due, these gentlemen give a smattering of attention to the rest of Europe and proceed to ignore the rest of the world. The Jensenites might try to clear this up in some way. They might even look into intra-group differences within the top ten. I would suspect that many would be found and that it would be healthy to make this known at professional meetings, in the journals and in the news media.

We also have a religious wing in this group who suspect that English-American children who are brought up in Southern Baptist churches perceive things differently and might really deserve the top spot upon the pyramid. Southern-

English-American-Episcopalians regard these assertions with a great deal of amusement. But who really knows? We all will if the ethnic learning line of research is extended logically to include every possible ethnic, regional and religious stock.

Also in the status research vein, we need research on the effects of racism and caste status on learning. The Jensenites can provide this by following Robert Coles and others around in Mississippi and South Carolina to study the parasitic worm and starvation situation among black children. Autopsies of a few who died might yield valuable evidence on the brain damages wrought by malnourishment. The team could change themselves into black people ala John Griffin and run the hostility gauntlet as they tried to find some information in the local library. Or the hilarity gauntlet as they made application for a professional or skilled job. They could fly as black men to Boston or Oakland and make the same applications to the craft union nearest the airport. Or they could try to get a tenured appointment in the Harvard Graduate School of Education, or a spot on the HER editorial board, or simply a rank higher than assistant professor among the 7,000 member Harvard faculty.

The Jensenites could give the same black injections to their children, enroll them in a different school and record what happens to them. Children learn efficiently if listening, reading, discussion, peer-group interaction, library resources and teacher-pupil interaction are all used efficiently. The investigators might be very interested in the change in quality in the last four areas for their now black off-spring and to see who is to blame and how the situation can be improved. To add a spicy dimension, low IQ scores could be substituted in the transfer folders.

Creation of multi-ethnic and multi-racial tests would also be a method of bringing some good out of the situation. If the only way to make *exactly* the same score on test items is to be of the same race, economic class, ethnic stock, and religious persuasion as the committee that developed the instrument, then we either must make intensive efforts to inter-marry, re-distribute income and institute religious purges and programs in this country or we must try to integrate more multi-racial and multi-ethnic material into the instruments. Said in the words of Dr. Nathan Wright, the Newark black power theorist, we must try to "dehonkify" the instruments.

Or we might decide that making *exactly* the same score is not important for all races and religions and come up with an Ethnic Success Quotient for tests based on validation studies of all of the hypenated groups we are going to study. Under such a system a Richmond born, Episcopalian, of English stock, from a family with an income of $12,000 would be declared below average if his Binet score was below 120. A score of 100 would relegate him to success quotient oblivion as a low normal. The Beaufort County, S. C. black children with worms might have a success quotient of 90 based on performance of adults from this sort of situation who somehow scrambled up the ladder. A black 100 score in this county would indicate a ESQ of potential genius.

Finally, in this vein, the Jensenites might make their most important contribution if they could somehow join with Earl Schaefer of the National Institutes of Health and others at the Universities of Florida, Western Michigan, etc. who are fastening on early infant stimulation and teaching as the key to agility on standardized tests. (The problem, of course, may be in getting the Schaeferites to join with the Jensenites given the Klan types who have embraced the latter as their own). Schaefer has already published some fine results of efforts with black children. The logic here is simple and very much in the vein of Cronbach's rebuttal to the Jensen paper, i.e., if you want black kids to think like white kids, imprint this type of thinking habit early (5 days to 2 years of age) with simple thinking, concept cluster tasks. White teachers can enable black parents to learn how. White disadvantaged children are being imprinted in the same manner in some studies. Ethnic and religious backgrounds have not been treated as yet. There might be a problem or two here regarding people who might want to imprint their children with their own brand of thinking or who have deep affection and preference for certain racial, ethnic or religious ways of thinking. Other parents might not want the new imprints to attend their schools on an integrated basis or live in their neighborhoods and play in their recreation centers. Something in the imprinting would thus be lost in this sort of forced isolation. But I am certain these reservations can be swept aside in the name of psychological research and the cognitive homogenizing process can progress.

Now for a closer look at some of Jensen's theories about black IQ. To begin, I received a form letter from Jensen in response to a request for clarification of his *real* stand on the implications of racial genetic inferiority that seemed to shine through the somewhat hazy statements of conclusion of his paper at the AERA, implications which the press quickly translated into flat statements of white intellectual supremacy. His article was based on this paper and gave the same impression to the press. (See Joseph Alsop, *Washington Post*, March 11; *Virginian-Pilot*, March 12: "Yet there is no use being mealy-mouthed about it. Dr. Jensen is really saying that in *addition* to the handicaps wickedly imposed by prejudice and discrimination, the average black American begins the race of life with a detectable genetic handicap").

Jensen's letter was addressed to the *Berkeley Daily Gazette* which he feels mis-interpreted his position. The following are excerpts from the letter:

Obvious differences in inborn mental ability 'between races'—these are a reporter's words. They certainly are not mine. The quotation marks, attributing this phrase to me, are therefore wrong. Furthermore, the statement is quite indefensible. The complex *causes* of objectively measurable differences in mental abilities among individuals or between different socioeconomic and racial groups are not at all 'obvious'.

Although my study of the existing evidence has led me to the position that intelligence differences among individuals, social classes, and racial groups are conditioned by both genetic and environmental factors, the estimation of the relative contributions of these influences is a problem of great technical and practical difficulty for researchers in behavioral genetics, and the research so far has been inadequate as a basis for definitive conclusions about racial differences in intelligence.

Jensen's treatment of the racial aspects of IQ in his article comes to the same point of inconclusiveness. It is very, very unfortunate that he, or the editors, failed to include a clear statement to this effect. Truth squad operations such as this letter and the rebuttals by psychologists in the HER Spring issue will never get read.

Jensen's second error in my estimation was to lean heavily on the Coleman Report for data on black inferiority. This report has been heavily criticized for inaccuracy. The most notable criticism is contained in the Winter, 1968 issue of the Journal of Human Resources in an article by Bowles and Levin. Sampling procedures, lack of cooperation by big school systems, failures to match black-white sample by curriculums, over-reliance on administrators' contentions that black-white facilities were indeed separate but equal (black parents in Eutaw, Alabama must have thought the research team had been smoking pot when they read the conclusions of the report) and crudeness of statistical measures were all analyzed as weaknesses which, when added to the fact that the study was made in pre-ESEA days, relegated it to the status of a 737 page, million dollar pilot study. On page 292 of the report, the authors state similar disclaimers, especially regarding the precautions necessary in interpreting their statistics.

In regarding as law this report's conclusions that the average black kid can get no further than a 9th grade operating level after 12 years of public school, Jensen ignores completely (or is unaware of) the record being compiled by the JOBS program of the National Alliance for Businessmen. These gentlemen take black drop-outs, place them on the job half-time and in reading and math classes half-time; they produce a two-year gain on tests every six weeks.

Jensen's major error, I believe, was his inconsistency in following a definite line of reasoning regarding the separation of gene linkage and pre-postnatal ravages of protein malnutrition. The latter is the most intensively researched thesis these days with NIH teams leading the way. Jensen did not even mention this line of research which (together with research in infant stimulation) I believe has answers for 42% mental retardation found in low-low (Jensen's level V) income black children and a lot of the other differences. In a half-starved brain like these kids have, how are we to really know if high or low IQ genes were linked? Jensen did not tell us how.

Jensen calls compensatory education a failure. So did reporters of the *Washington Post* who in turn received and printed a report by the ESEA staff of the Virginia Department of Education calling their allegations inaccurate and stating that they had hard data to back their claims. In response to a request for same, I received tables for statewide pre-post testing of 10,200 pupils in 15 school districts for 1967-68. The data show average month's increase in grade equivalency per month of 1.06 of instruction or an average overgain in achievement of more than a half a year per pupil as a result of compensatory education. Children scoring in the lowest decile had decreased from 41% to 28%. In the second quartile the number jumped from 8% to 16% and the drop-out rate had decreased by 63%. The officials noted that age-grade decrement had been scotched and that they believe that they had convincing evidence that their Title I program was a success. And this from one of the more conservative states in the Union and one with a record of slow starts in educational innovations. School people, it seems, are just now learning how to run compensatory programs. Or really try to. The first report to the President of the National Advisory Council on Disadvantaged Children noted this reluctance to really plan and implement on the part of many school systems. They quoted one superintendent who stated flatly that "it was useless and a waste of money to teach those jigs anything." Let us all hope he has since initiated a good program and that he doesn't read Jensen's article.

In drawing conclusions from 200-300 comparative studies of black-white IQ, Jensen failed to consider that all of the pre-1948 studies and most of the post-'48 studies failed to give attention to the deprivation axioms made popular by the University of Chicago group (Davis, Eels, et al) and until recently almost no psychometrists gave attention to the fact that white examiners in a black classroom are, in many, many cases, getting an invalid test performance. Their color, voice, manner, gestures turn many kids off, and they refuse to try. *This phenomenon is growing in intensity and must be dealt with.* How are you going to have a valid test session with kids who read in black papers and magazines that white researchers are sending their kids to Harvard by over-studying the black communities with federal grants? Or with kids who received a leaflet from a community group blasting tests as an "unfair tool of colonialists who control the black community"?

I believe that Jensen is wrong and I hope he does not do too much damage. I believe the HER editorial board should publish the rebuttals in the same issue with future attacks on the Negro. Rumors abound that attacks on the Negro church are planned. This will scotch the sensationalism of the press caused by the lag in time between issues. Indeed, the rebuttals will never be read by reporters, much less printed.

Jensen failed to take into consideration the black infant mortality rate as a factor in black infant supremacy on the motoric area of the Bayley Scales. This rate is three times that of white infants. Black kids must literally undergo a survival of the fittest test to be born, once conceived, and to stay alive.

Jensen has a serious contradiction in his analysis of tests and studies of black IQ. After offering half dozen or so studies to document his thesis that black kids don't do as well on IQ tests as white kids, Jensen closes his paper by stating that IQ tests fail to measure the full potential of black kids.

Jensen failed to consider the 1969 report of the Research and Evaluation Branch of Project Head Start in writing off Head Start gains as transitory. According to this report of several studies of the maintenance of gains, the investigators concluded that the gains were maintained when the children were enrolled in first grades or kindergartens in middle-class schools. Edmund Gordon of Teachers College and John McDavid of Miami led the team which wrote this report.

Jensen, like other psychologists, is completely incapable of un-raveling what would have to be un-raveled in order to separate genetic from environmental influences where American black and white people are concerned, to wit:

1. If 90% of the black people in America have ancestors that include white people, how can we tell when black genes or white genes make for a wrong mark on a test score sheet?
2. If a large per cent of white people have black ancestors, who are they? Are their samples controlled for this factor? Which genes, black or white, make for right marks on a test score sheet?
3. How can we parse out the effects of brain damage, brain stunting (due to malnutrition) and lack of early stimulation? Which accounts for a wrong mark on the test score sheet?

4. How can we parse and measure the degree of access and *welcome* of black people to cultural learnings?
5. How can we parse and measure the interest in and acceptance of the white "way of life" by black mothers and children? One can't get good scores on a "way of life" test like IQ unless one lives and accepts this life fully.
6. How can we develop indices which show comparability of school strengths, weaknesses and emphases? The school assessment study by Tyler's group is just getting underway over loud cries from many school people.

Jensen failed to consider the learning styles of black parents and the origins of these learning styles when he made white-black comparisons on associative and problem-solving learning. If you go to many rural schools in the south today, you will find the associative type of learning proceeding as it has for many, many years —*for both races*. This is the learning heritage of most big city black parents. They pass this style on to the kids early and it shows up in test profiles. If conceptual learning is viewed as a gradual acculturation process and offered early in school careers, these kids can be made to think. Jensen's exhortations to teachers to rely completely on associative learning might preclude this ever becoming a reality, however. Before any more articles are published, I think Jensen should do more work in the area of black history, demography and culture and that he should try to get into the area of racism and isolation and the big role they play in differences. There really is merit in his actually taking the black injections and getting first-hand information. He would only have to be a black man for two months.

Jensen's "g factor", the main basis of his claims for white supremacy, cannot be accepted as the mysterious phenomenon he postulates. Even little children now know from their television science that if something really exists, scientists will isolate it and measure it—especially before making serious conclusions about it.

I believe Jensen made two good points. One is that IQ tests don't show the full learning potential of kids who are poor and black. I was happy to learn that he had invented a test which does a better job. We should all buy it. He should make millions. The other is that intensive instruction rather than "cultural enrichment" is necessary to make these kids learn if they are locked in neighborhood schools. Unlike Jensen, I believe that they can proceed from associative learning to abstract reasoning if the instruction gradually brings them to this point. And even with this, I believe black kids will continue to think and score test items differently until full equality is achieved. Black kids screen out much of the curriculum and perceive the rest differently. Consider perceptions of Tarzan and the British Empire, for examples. Of course some black nationalists feel that it is a blessing that black people don't think like white people. As long as they can handle modern technology, make war, manipulate stocks, etc., I don't guess it really matters.

I believe the most potent strategy in the end will prove to be a combination of early stimulation and imprinting, and integrated schools with teachers who are free of racial and social class prejudices. IQ tests will also be eliminated from the schools. This is the strategy on which Neil Sullivan based his cross-bussing operations for the Berkeley schools. This may account for some of Jensen's concerns and reservations and perhaps, for his article. Pettigrew and others presented evidence in their work for the Civil Rights Commission that the earlier black children were placed in integrated schools, the closer they came to white norms on achievements tests. In turn, the white children came closer to perfection in their social learnings while losing no ground in test proficiency. The black children pick up the mysteries of Jensen's "g-factor" through association, I suppose, while the white children pick up the mysteries of "soul."

Reducing the Heredity-Environment Uncertainty

ARTHUR R. JENSEN
University of California, Berkeley

In his comments on the seven responses Dr. Jensen replies to criticisms, and suggests some appropriate research endeavors that could provide answers to the questions raised in his original article.

When the Editors of the *Harvard Educational Review* invited me to write a comprehensive summary of my research and thinking on the subject of educationally relevant individual differences, with reference especially to their genetic basis, I was delighted for the opportunity to present my views to the diverse and sophisticated audience that is reached by this journal.

One of my purposes in writing "How Much Can We Boost IQ and Scholastic Achievement?" was to provoke discussion among qualified persons of some important issues I believe have been relatively neglected in our common concern with improving the education of children called disadvantaged. Therefore it is a source of great satisfaction to me that the Editors have solicited and received extensive discussions of my article from several distinguished psychologists and an eminent geneticist—men whose own research in a variety of fields most germane to the contents of my article is widely known and highly respected.

Points of Agreement

It is of interest that many of the reports of my article in the public press have tried to make it look as though the several commentaries solicited by the Editors are strongly opposed to my paper and are in marked disagreement with its main points.[1]

---

[1] *U. S. News & World Report* (March 10, 1969), *Newsweek* (March 31, 1969), *Science News* (April 5, 1969), *Time* (April 11, 1969).

Harvard Educational Review Vol. 39 No. 3 Summer 1969, 449–483

In fact, seldom in my experience of reading the psychological literature have I seen the discussants of a supposedly "controversial" article (in the Editors' words) so much in agreement with all the main points of the article they were asked especially to criticize. On my main points the discussants agree with me at least as much as they agree among themselves, which is considerably.

The Role of Heredity

On this central theme there is essential agreement. Crow, the population geneticist, states: "That the heritability [of intelligence] is large is a justifiable conclusion at this stage. . ." "I agree with Jensen in deploring an uncritical assumption that only environmental factors are important and that genetic differences are negligible." "We should also realize that to whatever extent society is successful in its goals of providing equality of opportunity, to that extent the heritability [of mental abilities] will increase." Bereiter, a leader in psychometrics and in early childhood education, makes the same points: "The heritability of intelligence is unquestionably high, but what is more to the point is that with further social progress the consequences of heredity can only be more important because of the elimination of such sources of environmental variance as differences in the quality of education, nutrition, and medical care." Cronbach, our most eminent educational psychologist, says there is no doubt that "performance—intellectual, physical, or social— is developed from a genotypic, inherited base." Elkind, the leading American exponent of Piagetian psychology, emphasizes Piaget's agreement with genetic and biological maturational factors in cognitive development. Piaget's indices of cognitive development, such as the ability to conserve quantity, area, and volume, have been factor analyzed along with traditional psychometric measures of intelligence and are found to be highly loaded on the g (general intelligence) factor (Vernon, 1965); and Tuddenham (1968) has found social class and racial differences on a psychometrized form of the Piagetian developmental tasks that are comparable to those found for nonverbal IQ tests. Other supporting evidence relevant to this conclusion has been reviewed by Kohlberg (1968) in a paper highly germane to my own formulations. An interesting indication of the role of genetic factors in these Piagetian indices of cognitive development has recently come to my attention in a study by De Lemos (1966), who found that a majority of the full-blooded Australian aborigines who were examined on a variety of Piagetian conservation tests still did not show conservation of quantity, weight, volume, number, and area, even by the time they had reached adolescence. (The majority of European children pass these tests by seven years of age.)

These tests were passed, however, by a significantly larger proportion of aboriginal children who had one European grandparent or great-grandparent. De Lemos does not account for these results in terms of possibly differential environments. De Lemos's data are shown in Table 1.

TABLE 1

Numbers of Full-Blood and Part-Blood Australian Aboriginal Children Passing Piagetian Conservation Tests and the Significance Level (p) of the Difference[a]

	Age 8 to 11 Years			Age 12 to 15 Years		
	Full	*Part*	*p*	*Full*	*Part*	*p*
Total $N =$	25	17		17	21	
Tests						
Quantity	2	6	<0.1	2	15	<0.01
Weight	9	11	<0.1	7	17	<0.01
Volume	0	5	<0.05	2	4	N.S.
Length	10	10	N.S.	3	13	<0.05
Number	0	4	<0.05	3	8	N.S.
Area	1	4	N.S.	2	8	N.S.

[a] Source: De Lemos (1966).

Genetic Component in Race Differences

Here, too, there is considerable agreement, although it is qualified in some instances in ways that I will examine in later sections. In my paper I proposed simply that the hypothesis of genetic racial differences in mental abilities is a reasonable one deserving of further scientific investigation. Crow states: "I agree that it is foolish to deny the possibility of significant genetic differences between races. Since races are characterized by different gene frequencies, there is no reason to think that genes for behavioral traits are different in this regard." Cronbach agrees that "the genetic populations we call races no doubt have different distributions of whatever genes influence psychological processes." He then goes on to say: "We are in no position to guess, however, which pools are 'inferior.'" On this statement two comments are in order: First, who has advocated that we merely *"guess"* about racial genetic differences? I am advocating that we seek objective answers regarding genetic differences through appropriate scientific research. Again, the point I made in my article was that the present evidence on this topic is such that the hypothesis of genetic racial differences in intelligence is not an unreasonable one and should therefore be the subject of scientific investigation. Second, why does Cronbach put quotation marks around the word *inferior?* Lest the reader incorrectly infer that Cronbach is quoting me, let me note that I myself do not use this term and I object to it in this general context. I have said that there are racial and social-class differences in *patterns* of abilities and that there are probably genetic as well as environmental factors involved in these differences. The terms *inferior, superior, high, low, above, below,* etc. are meaningless in psychological discussions unless some particular dimension in the whole realm of abilities or traits is clearly specified and its relevance to a particular environmental adaptation is understood. Cronbach knows as well as I that it is nonsense to speak of different racial gene pools in general as *superior* or *inferior.*

Possible Dysgenic Trends in Our Population

In my paper I raised the question: "Is there a danger that current welfare policies, unaided by eugenic foresight, could lead to the genetic enslavement of a substantial segment of our population?" Differential birthrates in the population that are correlated with educationally and occupationally relevant traits of high heritability could produce long-term dysgenic trends which would make environmental amelioration of the plight of the disadvantaged increasingly difficult.[2] Hunt, psychology's most eloquent and influential spokesman for environmental amelioration of educational handicaps, states that ". . .the national welfare policies we established in the 1930's have probably operated in dysgenic fashion, and that it is highly important to establish welfare policies which will encourage initiative and probably, in consequence, help foster positive genotype selection." Hunt points out how some social and educational programs, such as involving parents in programs of early childhood education, can produce not only direct benefits to the children enrolled in the program but also more indirect benefits to the future welfare of the families involved, as when parents voluntarily enrolled in a Planned-Parenthood clinic. Says Hunt: "The enrolling in the Planned-Parenthood clinic suggests that this kind of enterprise in early childhood education instigates help to prevent some of the dysgenic processes with which Professor Jensen and I are both concerned. Hunt also agrees that it is "highly important to raise the intelligence, the educational

[2] For instance, unless existing trends markedly change, it can be predicted that within the next 20 years more than a million children with IQ's below 70 will grow up in fatherless homes in our urban slums. The amount of human frustration and suffering implied by this prediction, if it becomes reality, is incalculable.

attainments, and/or the general competence of those people who now comprise the bottom quarter of our population in measures of this cluster of characteristics."

If Hunt believes there have probably existed dysgenic trends in some segments of the population since the 1930's, he must logically conclude that he also believes there are heritable behavioral differences among some segments in the population, socially and educationally relevant behavioral differences that exist within every racial group, although he does not say this explicitly. There is, of course, nothing "inevitable" about these genetic differences in the sense of their being predestined or immutable or inherently associated with race *per se*. Whatever they are, if they indeed exist, they are undoubtedly a product of differing historical, social, and environmental selective pressures. The really important point now is to try and understand the genetic trends in the population resulting from current social forces, and if dysgenic trends indeed exist, to discover the kinds of social conditions and public policies that can be created in a humane, democratic society to counteract and reverse such trends for the good of all, especially of the generations not yet born.

Value of Compensatory Education Programs

I am essentially in agreement with Hunt's evaluation of the failures of compensatory early childhood education and the reasons for the ineffectiveness of preschool programs based on the free-play socialization model of the traditional nursery school. One must also agree with Hunt that we cannot now evaluate forms of compensatory education that have not yet been tried or even invented. The fact remains, however, that our most massive, large-scale attempts at what has been called compensatory education have apparently not produced the desired or promised results. I cited the comprehensive evaluation of the U. S. Commission on Civil Rights (1967), which arrived at this negative conclusion after a nationwide survey of the major Federally-funded compensatory programs. I favor continuing experimentation in improving the education of the disadvantaged, and I favor trying a wide diversity of reasonable approaches. In our present state of ignorance about how best to teach children who are spread over an enormously wide range of abilities and proclivities and diverse cultural backgrounds, we are hardly justified in launching nationwide compensatory programs of massive uniformity. The same expenditures invested in a real *variety* of smaller-scale programs that psychologists, educators, and parents have some reason to believe might succeed, and which can be properly evaluated, will more surely and quickly lead to knowledge of which policies and practices will or will not produce the most beneficial results. We *have* learned from many of the programs evaluated by the U. S. Commission on Civil Rights what kinds of measures have produced no signs of success, though they have been put to the test for from three to eight years. It is a half-truth to say that these programs have not had a fair trial. Thirty years after the beginning of the progressive education movement, its extreme proponents, then on the defensive, were still saying it could not be evaluated because it had not been tried for a sufficient time. At least from the evidence now at hand, I must agree with Cronbach's statement that there has been "too much blithe optimism about our ability to improve the intellectual functioning of the slum child and the retarded child." And Elkind says "What is the evidence that preschool instruction has lasting effects upon mental growth and development? The answer is, in brief, that there is none." Bereiter, on the other hand, presents new evidence from his own excellent work with disadvantagd preschool children showing substantial gains in intellectual skills resulting from specific forms of intensive instruction. These are exciting findings and we will want to follow this work closely in the future. The crucial question, we all recognize, still concerns the permanence of the gains and the factors that affect their durability. The answer is still in the future.

Points of Disagreement

The points of disagreement seem to me less fundamental and much narrower in scope than the points of agreement. Some of the most critical-sounding statements quoted so repeatedly in the public press actually have little if any substance to back them up when read in context. At least two of the discussants seem to disagree with each other regarding my objectivity and accuracy. Crow states: "Jensen's article, together with many others that he has written recently on this subject . . . , constitutes a thorough review and synthesis of the various attempts to apply these methods [of biometrical genetics] to human intelligence and scholastic achievement. Jensen has become a leader in this field, and I, as a population geneticist, admire his understanding of the methods and his diligence and objectivity in bringing together evidence from diverse sources. He presents the evidence fairly, relying on empirical data in preference to introspection or traditional wisdom, and is very careful to distinguish between observation and speculation." Cronbach, on the other hand, makes a highly contrasting statement in the first paragraph of his paper: "Unfortunately, Dr. Jensen has girded himself for a holy war against 'environmentalists,' and his zeal leads him into over-statements and misstatements." Since this has become the most widely quoted critical statement in the press about my article, I would like to examine it.

Let readers judge for themselves if there is anything warlike about my article. There is little doubt, however, that in recent years students of the behavioral and social sciences, educators, and the public in general have been strongly propagandized with the views espoused by extreme environmentalists, and that these views have become a basis for official policies.[3] If Cronbach interprets my confronting those he refers to as "environmentalists" with some of the scientifically-ascertained facts concerning the genetic aspects of mental abilities as being a "holy war," that is interesting in itself. What Cronbach calls a "holy war" I call simply looking for the facts.

But what about the more serious allegation that Cronbach goes on to make— that of "over-statements and misstatements" in my article? Cronbach does not follow up on this charge. He does not point to a single example of an "over-statement" or a "misstatement" in my paper. The closest Cronbach comes to indicating specifically what he might have had in mind in using these words is later on, where he says: "I have detected substantial distortions in Jensen's report of some research, and I must therefore warn the reader against accepting his summaries. Selective breeding studies are a case in point . . ." Let's take a close look at how Cronbach follows up on this attempted broadside.

Selective Breeding Studies

I stated that rats can be bred for maze-learning ability. I also pointed out that maze learning is a complex behavior, involving a host of sensory, motor, temperamental, neurological and biochemical components. Nevertheless, the molar behavior of speed of learning to run through a maze without entering blind alleys, I said, can be selectively bred. Cronbach seemingly challenges my statement by pointing out almost exactly what I had already stated in my own paper, namely, that maze-learning ability is a result of many factors. One can breed for any particular pattern of these factors, depending on the nature of the learning task and the criterion which serves as the basis for selection in the breeding of successive generations. Cronbach notes that the Tryon strains were bred to one kind of maze

---

[3] We find, for example, a statement from the U. S. Office of Education (1966): "It is a demonstrable fact that the talent pool in any one ethnic group is substantially the same as that in any other ethnic group." And from a Department of Labor (1965) report: "Intelligence potential is distributed among Negro infants in the same proportion and pattern as among Icelanders or Chinese, or any other group." There is simply no factual basis for these official pronouncements, which I believe are motivated more by political than by scientific considerations.

under one kind of incentive. Is the selective breeding for maze learning in one highly specific set of conditions any *less* genetic than breeding for maze learning ability that generalizes across many different mazes? In fact, in the study which I cited as an example, and from which my Figure 4 is taken, rats were bred for learning ability that generalized across 24 different mazes. I would call this a fairly general factor of maze learning ability. Fuller and Thompson (1960) in their well-known textbook, *Behavior Genetics,* say of this experiment:

Thus a fairly broad range of rat intelligence was sampled. The procedure involved a lengthy period of habituation for all animals on simple pretest problems until a certain criterion was reached. In this way, the influence of motivational and emotional differences was minimized. The Hebb-Williams maze, generally speaking, is analogous to human intelligence tests which involve a large number of short items usually administered only to subjects who have had previous preparation. (pp. 212-213)

Since the 1953 paper by John Paul Scott that Cronbach refers to as an "eloquent attack on the idea of a general inherited learning ability" predates the Thompson experiment to which I referred, the only maze learning experiments it cites being those by Tryon, who bred rats for a specific maze ability, it can no longer be regarded as an adequate account of what we now know about selective breeding for maze-learning ability. Indeed, I have found no evidence in the literature of a general learning ability factor in animals that generalizes across a wide variety of different *types* of learning. But this fact is actually irrelevant to the question of a general factor in human intelligence, which we know to have a large genetic component and would therefore unquestionably respond to selection. Cronbach concludes this section by saying: "Jensen cites Scott as if he endorsed such an idea" [of a general learning ability in animals]. I did no such thing. As readers of my article can plainly see, I cited Scott & Fuller (*Genetics and the Social Behavior of the Dog,* 1965) along with Fuller & Thompson (1960) strictly in connection with my general introductory statement to this section, to the effect that behavioral traits respond to selective breeding in animal experiments. These are still the best two general references I can give for this statement.

Twin Studies

Kagan, a leading developmental psychologist, similarly criticizes parts of my paper in a way that hardly stands up under close examination. For example, he cites Gottesman, a behavioral geneticist, as questioning "the validity of Jensen's ideas." From Gottesman's article (1968, p. 28) Kagan reports: "In a study of 38 pairs of identical twins reared in *different environments,* the average difference in IQ for these identical twins was 14 points, and at least one quarter of the identical pairs of twins reared in different environments had differences in IQ scores *that were larger than 16 points.*" Gottesman, however, provided a bit more information. Actually *two* intelligence tests were used: a vocabulary test and a nonverbal test of abstract reasoning. The vocabulary test showed the average twin-pair difference of 14 points; the nonverbal test showed a difference of 10 points. Kagan himself italicized *"different environments,"* so let us look at the average difference on these tests between twins *reared together*: vocabulary = 9 IQ points, nonverbal = 9 IQ points. The average difference between the scores of the *same persons* tested twice on the same tests can be inferred from the reliabilities of these tests: vocabulary = 4 IQ points, nonverbal = 6 IQ points.[4] But the best way of seeing whether the Gottesman review cited by Kagan "questions the validity of Jensen's ideas" is to look at the original study which Gottesman summarized, which is one of the most careful and rigorous twin studies ever conducted (Shields, 1962). Shields' twin correlations are shown in Table 2. I ask, do these results "question the validity" of any of the statements in my article regarding the heritability of intelligence? To go on to say, as Kagan does, that the difference between members of identical twin pairs reared apart is larger than the average difference between black and white populations finds absolutely no support in this evidence! Kagan does not mention the statistical fact that the average absolute difference between twins includes the tests' measurement error, while the difference between the means of large groups does not contain this source of error.* The average absolute differences for height, intelligence, and scholastic achievement between a variety of kinships are shown in Figure 1.

In a similar vein of criticism is Hunt's comment: ". . . it is interesting to note what he [Jensen] omits from a paragraph quoted from the geneticist Dobzhansky," whom I quoted in part and paraphrased in part. Hunt's statement implies that the part of Dobzhansky I did not directly quote contradicts my own views. The omitted portion of Dobzhansky reads: "Although the genetically-guaranteed educability of our species makes most individuals trainable for most

---

[4] The standard error of measurement of most IQ tests is between 5 and 10 IQ points. This source of error is estimated by testing the same person twice or from split-half scores of odd vs. even numbered items.

TABLE 2

Correlations Between MZ Twins Reared Together and Apart[a]

Measure	Twins Reared Apart (N = 44) r	Twins Reared Together (N = 44) r
Mill Hill Vocabulary	.74	.74
D48 Domino Test	.76	.71
Composite Intelligence Test Score*	.77	.76
Composite Intelligence Test Score Corrected for Attenuation	.86	.84
Height { Males	.82	.98
Height { Females	.82	.94
Weight { Males	.87	.79
Weight { Females	.87	.81
Extraversion**	.61	.42
Neuroticism**	.53	.38

[a] Source: Shields (1962), p. 69.

* Mill Hill Vocabulary Scale and the D48 (Domino) Test

** Maudsley Personality Inventory

occupations, it is highly probable that individuals have more genetic adaptability to some occupations than to others. Although almost everybody could become, if properly brought up and trained, a fairly competent farmer, or a craftsman of some sort, or a soldier, sailor, tradesman, teacher, or priest, certain ones would be more easily trainable to be soldiers and others to be teachers, for instance. It is even more probable that only a relatively few individuals would have the genetic wherewithal for certain highly specialized professions, such as musician, or singer, or poet, or high achievement in sports or wisdom or leadership." The reader can see for himself if Dobzhansky's statement in any way contradicts my own paraphrase.[5]

[5] The paraphrase read: "Some minimal level of ability is required for learning most skills. But while you can teach almost anyone to play chess, or the piano, or to conduct an orchestra, or to write prose, you cannot teach everyone to be a Capablanca, a Paderewski, a Toscanini, or a Bernard Shaw."

FIGURE 1.

Correlations, r, (corrected for attenuation, i.e., error of measurement) between persons with different degrees of kinship and reared together or apart. The average absolute difference (corrected for error of measurement) between pairs of individuals is based on the same scale for height, intelligence, and scholastic achievement, with a standard deviation (SD) of 16, the SD of Stanford-Binet IQ's in the normative population (Jensen, 1968a).

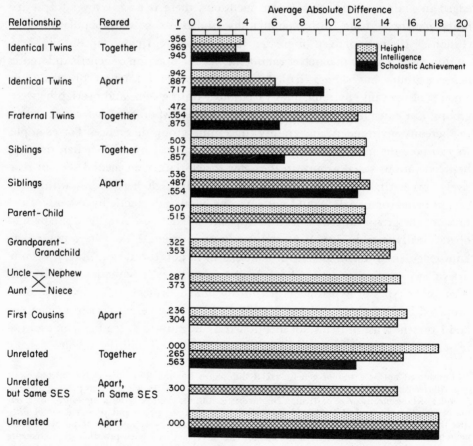

Individual Differences vs. Group Differences

Kagan further claims that my article contains "a pair of partially correct empirical generalizations wedded to a logically incorrect conclusion." The "partially correct" empirical generalizations he refers to are (a) the high heritability of intelligence (is there contrary evidence?) and (b) the average difference of about one standard deviation (15 or 16 IQ points) between Negro and white children on standardized intelligence tests (is there contrary evidence?). The "logically incorrect conclusion" is that, given these two facts, the IQ difference between Negro and white children, therefore, involves genetic as well as environmental factors. I have not drawn this "conclusion" from this premise, as geneticist Crow acknowledged in stating: "Strictly, as Jensen mentions, there is no carryover [of heritability measures] from within-population studies to between-population conclusions."[6] I have explained in greater detail elsewhere (Jensen, 1968b) that heritability coefficients by themselves cannot answer the question of genetic differences between groups, but when used along with additional information concerning the amount of relevant environmental variations within groups and overlap between groups, can enter into the formulation of testable hypotheses that could reduce the heredity-environment uncertainty concerning group differences. For example, we can pose the question: are differences (as measured by, say, median overlap) between various racial groups in the same society larger on mental tests of relatively low heritability than on tests of relatively high heritability within the groups being compared? Would not environmental and genetic hypotheses of the cause of the group difference lead to opposite predictions? Are these predictions operationally testable, just as other hypotheses in science? They have not, to my knowledge, been tested, and so, of course, I have not, contrary to Kagan's claim, drawn any conclusion about the outcome of such an hypothetical experiment. Also, other types of experiments permitting much stronger inference have been proposed but have not yet been done. I simply say there is sufficient evidence— and I present a list of items not mentioned by Kagan—to suggest it is not an unreasonable hypothesis that racial differences in mental abilities involve genetic

[6] Considered not as a test of genetic racial differences but merely as an abstract problem in quantitative genetics, I wonder if Crow would not agree with the following: Given two populations (1 and 2) whose means on a particular characteristic differ significantly by x amount, and given the heritability (H_1 and H_2) of the characteristic in each of the two populations, the *probability* that the two populations differ from one another genotypically as well as phenotypically is some monotonically increasing function of the magnitudes of H_1 and H_2. Such probabalistic statements are commonplace in all branches of science. It seems that only when we approach the question of genetic race differences do some geneticists talk as though only one or two probability values is possible, either 0 or 1. Scientific advancement in any field would be in a sorry state if this restriction were a universal rule. Would Crow argue, for example, that there is no difference in the *probability* that two groups differ genetically where H for the trait in question is .90 in each group as against the case where H is .10? In the absence of absolute certainty, are not probabalistic answers still preferable to complete ignorance?

as well as environmental and cultural factors. What factual or theoretical genetic evidence can Kagan present that this hypothesis is unreasonable or has already been scientifically rejected? Does Kagan advocate the fallacy that until a reasonable hypothesis has been definitely proved, we must believe that the *opposite* of the hypothesis is true? Or does he believe that these questions should not even be asked, much less formulated into testable hypotheses? My position is that reasonable hypotheses concerning socially and educationally relevant questions should be subjected to appropriate investigation and the findings be published and widely discussed by the scientific community and the general public as well.

The Bloom Fallacy

Cronbach notes that I refer to Benjamin Bloom's (1964) summary of age-to-age correlations of mental test scores up to 17 or 18 years of age. Cronbach believes that since I introduced this source I was also obligated to disclose that Bloom gives these data an interpretation opposite to mine. "Bloom sees the gains from year to year in test score as random and unpredictable, hence due to external events and not inheritance." I have no argument with Bloom's correlations, which are empirical fact. His interpretation of them, however, is fallacious, and though it does fit the correlation data themselves, it does *not* fit other data that are an essential part of the picture. These correlations, beginning at around zero between ages 1 and 18 years, gradually increase up to about .90 between ages 16 and 18. This pattern of correlations would result between series of scores if a number of random increments were added to each score starting with a base of zero (or some value without variance). But differences among the final scores, each consisting of the summation of random increments, will not be at all predictable. Yet we know that mental test scores are quite predictable, just from a knowledge of the parents' IQ's, even before the child is born. (The correlation of midparent and offspring at age 18 is about .70.) What the evidence on the heritability (*H*) of IQ tells us is that about 80 per cent of the variance in IQ's is conditioned by the genes, in other words, by factors already present at conception. This being the case, the interpretation of mental growth from birth to 18 years of age as a process of adding random increments just makes no sense. The Bloom model would be in accord both with the facts of the age-to-age correlations and with the facts of the heritability of IQ if it conceived of the adult level of ability as a genetically predicted level of ability from which random increments are subtracted, going in the backward direction toward birth. In other words, the genetic factors laid down at conception are increasingly realized in the individual's performance as he approaches the asymptote of that performance, in this case, ability on mental tests.

Cronbach also mentions late blooming in IQ, i.e., the fact that some persons show marked spurts in their relative position even as late as adolescence. Why should it be assumed that these mental growth spurts are environmentally caused? In fact, the relatively high correlation between identical twins across the whole age range, even in the range of the lowest year-to-year correlations, is a strong indication that genetic factors play a major part in the *form* of the individual's growth curve for intelligence, just as is true for height.

Underplaying the Role of Heredity

Cronbach says: "Jensen accuses writers on education of underplaying or denying the role of heredity. Some of this bias does exist, but Jensen is unfair. He does not quote the writers in psychology and education who do devote space to heredity." On the contrary, these are the ones about whom I have the greatest complaint. I do not criticize textbook writers who merely omit discussion of the heredity-environment issue. I *do* object to those textbook authors (Cronbach is *not* among them) who bring up the subject but then distort, misrepresent, or minimize the relevant evidence. I have recently surveyed 25 of the most widely used recent textbooks in educational psychology with reference to this topic and I am preparing a separate article on their treatment of the heredity-environment aspects of individual and group differences. Leaving out those few that say nothing about these topics, all but a few of the rest give what must be regarded as inaccurate or misleading information.

The Interval Scale of IQ

My argument that IQ's are approximately normally distributed in the population and that the IQ scale behaves like an *interval scale* is claimed by Hunt to be circular. Hunt shows that he misses the essential point when he says ". . . apparently, for Jensen, going twice around the circular argument removes its circularity?" The argument:

(a) We *postulate* that intelligence is normally distributed in the population, just as most other metrical biological characteristics (e.g., height, age of menarche, head circumference, etc.).

(b) We devise an intelligence test to yield a normal distribution of scores in a representative sample of the population. If intelligence is *in fact* normally distributed, and if our test scores yield a normal distribution, it necessarily follows that the test scores constitute an equal interval scale. (If the scale were transformed, as by taking the square, square-root, logarithm, or any other non-linear transformation of the scores, the distribution would no longer be normal.) So far the logic is, of course, circular, as is the first step in *all* forms of measurement in science.

(c) But then we go beyond the circularity by determining if our postulate (i.e., normality) and the system of measurement that is relevant to it (i.e., interval scale) can make quantitative predictions of some phenomenon which is itself entirely independent of our assumption about the scale of measurement. If the prediction is then borne out in fact, the circularity is broken. The independent phenomenon we wish to predict in this present case is the regression of IQ for different degrees of kinship. The amount of regression for quantitative traits for various degrees of kinship is predicted from principles of population genetics and holds for clearly inherited metrical physical characteristics which are definitely known to be measured on an interval scale (e.g., height)—and our method of measuring intelligence itself plays no part in these genetic principles or analogous physical traits, so we are no longer involved in a circular argument. The genetic predictions will be borne out, however, only if our measurements of intelligence constitute an interval scale, because the genetic predictions assume rectilinear regression lines between kinship for metrical traits. The fact that the obtained regression lines for IQ's are rectilinear and closely in accord with the predictions (the same predictions that would be made for height, head circumference, fingerprint ridges, etc.) means that the IQ measurements behave like an interval scale. The genetic evidence, reviewed in my paper, fully supports this. Make a nonlinear transformation of the IQ scale and what happens? The kinship regressions are then clearly not rectilinear and the obtained kinship correlations are not in accord with the genetically predicted values. Furthermore, there is nothing in this whole argument which suggests, as Hunt accuses me of implying, that the present IQ distribution "is fixed in human nature for all time or until selective breeding alters it." Here Hunt again sets up his favorite straw man—*"fixed* intelligence."

The Editors' introductory summary of Hunt's paper says that "He [Hunt] finds Jensen's claims about the high heritability of intelligence unsubstantiated." Yet I find in Hunt's paper nothing that challenges either the theory or the methods or the findings concerning the numerous studies of the heritability of intelligence which are summarized in my article! If one wishes to argue with the empirical finding of a heritability coefficient (H) of, say, 80% for intelligence (the average value of H for the studies reported in the literature), then one must fault those

FIGURE 2.

Comparison of what the distribution of IQ's theoretically would be if all geno-types were identical (for IQ 100) in an "average" environment (assuming a normal distribution of environmental advantages) and all variance were due only to non-genetic (environmental) factors (heavy line). Under these conditions the heritability (H) of IQ's would be zero, instead of .80 as in the present population. The shaded curve represents the normal distribution of IQ's in the present population.

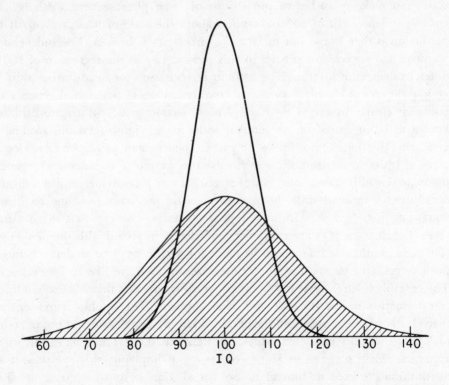

heritability studies which yield these results. Neither Hunt nor any of the other discussants has done this.

Phenotypic Variation of a Given Genotype

I wish to make it as clear as I know how just what a heritability (*H*) value of .80 actually means. Crow and Cronbach essentially reiterate what I said about the meaning of *H*. The latter says: "The index of .80 is impressive, but it is less dis-couraging than Jensen implies," and he presents a rather complex statistical argu-

FIGURE 3.

The theoretical distribution of IQ's if all variance due to environmental factors were eliminated (with everyone having an "average" environment) and all the remaining variance were due only to genetic factors (heavy line). Under these conditions the heritability (H) of IQ's would be 1.00. The shaded curve represents the normal distribution of IQ's in the present population, in which H = .80.

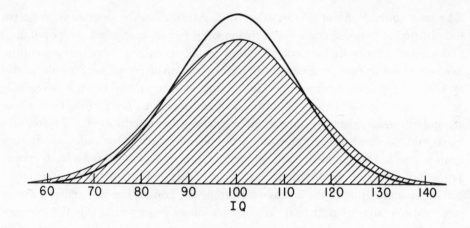

ment to indicate the range of phenotypic variation for a given single genotype which is implied by an *H* index of .80. The same argument·can be illustrated perhaps more simply by graphical means. I did this in my original manuscript, but it was edited out, probably because it seemed redundant. But I think the graphical explanation is worth the space it takes. Figure 2 shows the normal distribution of IQ's in the population (shaded curve), and the heavy-line curve shows the hypothetical distribution of IQ's if all persons in the population had exactly the same genotype for intelligence and the only sources of variation were environmental. The area under both curves is the same, but the tall curve has only 20% of the variance (i.e., 1 — H = .20) of the flat curve. In other words, it is the distribution of phenotypes for a particular genotype, given *H* = .80. This depicts essentially what Cronbach's statistical sortie was aimed to point out. But it is only *half* the picture. Figure 3 shows the reverse hypothetical situation, i.e., the difference in the IQ distribution (heavy-line curve) if genotypes remained as varied as they actually are but everyone had the same environment (pre- and post-natal), which, of course, is possible only theoretically. The population variance in IQ's is thus reduced by 20%, and Figure 3 is how it would look. To point to only one or the other figure alone is improper. It takes *both* to tell the true picture.

Points of Misunderstanding

Confusion Between Population Average and Individual Differences

The most common point of confusion among several of the discussants concerns the distinction between common environmental factors that affect the population average and factors that account for individual deviations from the population average. Genetic and environmental factors are involved in both of these two aspects (i.e., population mean vs. individual variation), though not necessarily to the same degree. If the population average were not susceptible to environmental influences, there would, of course, be no value in education! Children can learn and do learn when appropriate opportunities are provided, just as they grow when food is provided. And the average level of developed skills in the population will reflect to an important degree the extent and quality of the opportunities for learning, just as the average stature of the population will reflect to some degree the quality of nutrition. While widespread improvement in the environment relevant to a particular trait may raise the mean level of the population on that trait, it does not necessarily, or even usually, decrease differences among individuals. No one denies the importance of certain environmental conditions for the development of phenotypic characteristics. What heritability studies of intelligence show, however, is that in the European and North American Caucasian populations in which these studies were conducted environmental variations account for relatively little (about 20%) of the variation in intelligence among individuals. These studies by themselves can tell us nothing about changes in the mean of the population across generations. Even though the offspring may be brighter or taller than their parents, the *correlation* between parents and children does not change appreciably. For highly heritable traits, like intelligence, parental phenotypes thus remain a statistically reliable basis for predicting the deviations of their offspring from the population mean. Improving the population's relevant environment for the development of a trait usually *increases* the phenotypic manifestations of genotypic differences, and, as Bereiter points out, it *increases* the heritability of the trait: "One's view of the future beyond equality of opportunity must, therefore, be of a future in which differences in intelligence are virtually one hundred percent determined by heredity." Bereiter adds in a footnote: "This eventuality is in no wise to be forestalled by individualized instruction or any more libertarian tactic; on the contrary, such approaches should allow inherited differences to reach full flower, as advertised in the slogan, 'enabling each child to realize his fullest potential.' "

This brings us to the question of the primary aims of compensatory education. The aims are often explicitly stated as being to decrease or remove the scholastic (and ultimately occupational) achievement gap between children called disadvantaged and the rest of the population, or even to make all children perform at least at the population average for their grade level throughout their years in school. Educational innovations, improvements in instructional techniques, and so on, when they are successful, are just as likely to increase the learning and achievements of the advantaged as of the disadvantaged, with little if any decrease in the relative differences among individuals, so that E. L. Thorndike's dictum would remain valid: "In the actual race of life, which is not to get ahead but to get ahead of somebody, the chief determining factor is heredity." Equality of opportunity is a worthy and attainable goal. Equality of performance is a misguided hope. The important thing for the welfare of children and of society in general would seem to be to try and create conditions that will maximize the proportion of the population that can learn and work successfully and rewardingly in the diverse occupational roles that the society provides. It is clear that various peoples and societies in the past and in the present have approached this realistic goal to quite different degrees, and it would seem worthwhile to inquire into the social, biological, and educational conditions which have either hindered or promoted the realization of this goal. I would hypothesize that among the relevant conditions would be at least two prominent factors: (a) the working of eugenic pressures, either consciously and directly, or indirectly through the value system, social structure, socially-conditioned mating patterns, and the like, and (b) a wide diversity of educational options, paths, and goals.

Height as an Example

I have said that the mode of inheritance of intelligence quite closely parallels that for physical stature. Four of the discussants referred to the overall increase in height in the population as if this fact somehow diminished the importance of heredity in individual differences in height, and even more so in intelligence, since intelligence has a lower heritability than height (about .80 vs. .95). Because this has been one of the commonest arguments put forth by persons traditionally called environmentalists, I think it deserves a closer look than it was given by the discussants. The parallel between height and intelligence is close enough that we may gain some insights about the latter from a study of the former, about which much more is known concerning population trends across many generations.

Crow states that because of unidentified environmental influences height has increased by a "spectacular amount." And Hunt, on the basis of what he heard from guides at Jamestown's Festival Park and aboard the *U. S. Constitution,* states that height "appears to have increased *nearly a foot* without benefit of selective breeding or natural selection." Presumably Hunt is referring to the increase in adult height since about the 17th century. The implication is that all of this increase in height is strictly the effect of environmental and not genetic factors.

Let us see what more dependable authorities than tourist guides have to say about this subject. I have obtained my information from a book on human genetics by a noted British geneticist (Carter, 1962), and from comprehensive articles on this subject by J. M. Tanner (1965, 1968), the world's leading researcher on human growth. Here is what I find:

First of all, it is essential to distinguish between growth *rate* and final (adult) *level*. Adult height has increased little over the past century or so. Carter (p. 102) says that skeletal remains suggest there has been little appreciable change in height in Britain over the past 5000 years. "If there has been any increase [in adult height in Britain] it is only of the order of 1 inch. What environmental improvements appear to be doing is, in the main, to accelerate growth, so that full adult height is being reached earlier. Records from the armed services, prisons, and anthropological surveys suggest that full adult height has not changed by more than 1-½ inches for the past century" (p. 102). Other countries have shown slightly higher increases than in Britain, and Tanner (1968) concludes that adult height has increased 2-½ to 3-½ inches in the past century. Increases before the last century were relatively minute. While the increase in height since about 1700 was a positively accelerated curve, it has become negatively accelerated in the 20th century, and the trend is leveling off, especially in the United States. Growth *rate,* and consequently children's height, has shown much greater increases. Children now attain their full adult height by 18 or 19, on the average, rather than at 26, as was the case only 50 years ago. The trend toward earlier maturation shows up most dramatically in the lower age of menarche, or first menstrual period, which has declined from 17 to 13 years of age since 1840.

The trend toward earlier maturity seems to be related largely to environmental factors—probably improved nutrition and, it has been hypothesized, electric lights. (Children today spend more time awake and, due to electric lighting, more hours under illumination, so they grow for more hours per day, just as chickens raised under constant illumination reach egg-laying capacity much younger than when raised under normal conditions.) But part of the cause of increased growth rate is also genetic. The increase in *adult* height may be almost entirely attributable to genetic factors. Tanner (1965) points out that among environmental factors the increase in adult height is at least as closely related to the introduction of the bicycle and other improved modes of transportation as to improvements in nutrition and health care. What is the explanation? It is what geneticists call the outbreeding effect, heterosis or hybrid vigor. Tanner (1968) states that the "height of adults is significantly and inversely correlated with the degree of inbreeding in the region studied," and "the trend in adult height may have in whole or in part a genetic explanation." It has been estimated that 10 to 20 per cent of the variance in height is due to genetic dominance, so that the mean of the offspring of two parents will not be halfway between the parents but slightly closer to the taller parent. Outbreeding increases heterozygotes in the population with a consequent increase in height. This heterosis due to outbreeding also enhances growth rate and early maturation as amply demonstrated in numerous experiments in animal breeding. Outbreeding has increased at a steady rate ever since the introduction of the bicycle. For example, sons of parents who were from *different* Swiss villages were taller by approximately 1 inch than the sons of parents from the *same* village. Persons born to parents whose inbreeding is to the degree of first cousins average 1.4 inches shorter than persons whose parents are unrelated. According to Tanner, the average degree of outbreeding that has taken place in the last century can account for 0.8 inches increase in height per generation. The increase in heterozygosity, of course, eventually "saturates," and the effects level off, as has already occurred in the U. S. That genetic as well as nutritional factors are a major cause of the increase in actual height is further shown in the fact that approximately the same increase has occurred in all social classes in Western countries even though there have been nutritional differences among social classes. On the other hand, earlier maturation, as indexed by age of menarche, is more related to nutrition, as shown by a decrease in social class differences in countries with a very wide range of nutrition. Thus Hong Kong has shown a convergence between social classes in the decreasing age of menarche, while England and Scotland have not.

Have genetic differences between individuals and between groups *decreased* with the average increase in height in the population? No. Take the sex difference in height, which is surely genetic. Since males have responded more than females to improved nutrition, the sex difference in height has slightly increased. The range of individual differences in height is at least as great as ever it was and the heritability of height is probably higher than it has ever been.

Thus the slight increase in the population's mean height over the last two centuries—the environmentalists' favorite counter-argument to the high heritability of IQ—itself turns out to be largely a genetic phenomenon!

What has been said about height probably applies also to intelligence and other biologically-conditioned characteristics. There is some evidence, for example, of an increase in intelligence test performance in the general population between World War I and World War II (Tuddenham, 1948), due no doubt to improvements in education, nutrition and health care, and standards of living in general, and the same general factors involved in the increase in height. Intelligence variance, too, has a genetic dominance component not very different from height. Both white and Negro populations have shown the reported increase in intelligence test performance, but there has been no indication of a *convergence* of their mean scores since World War I, although there have been marked socioeconomic and educational advances since then. In fact, there is some indication from armed forces tests and nationwide testing surveys that, if anything, the average difference in performance between Negro and whites may have *increased* since World War I (e.g., Minor, 1957).[7]

Confusion of Cultural Disadvantage with Sensory Deprivation

Hunt's paper places great emphasis on the role of sensory stimulation in early development as a factor in later mental attainments. He cites particularly two classes of evidence in support of this hypothesis: (a) experiments on the effects of extreme sensory deprivation in animals, and (b) observations of children subjected in early infancy to extreme sensory deprivation and motor restriction through being confined in cribs in understaffed orphanages.

The connection between these lines of evidence and the average lower IQ's and deficiencies in scholastic performance of children called culturally disadvantaged is purely hypothetical. I seriously question the relevance of these types of evidence for understanding the observable abilities of disadvantaged children.

I do not contest the evidence showing that rabbits, kittens, and chimpanzees

[7] It has also been argued that our concern should be with the relative improvement of Negroes compared to the white population, rather than with the absolute improvement of one group. Though some differentials have been cut, a time-gap analysis indicates that the Negro lags about a quarter century behind the white, and this lag has not been reduced since World War I. On some measures, there is evidence that the environmental differences, expressed in time-lag, are increasing. See: Rashi Fein, "An Economic and Social Profile of the Negro American," *Daedalus*, 94, no. 4 (Fall, 1965), 815-846.

after being reared in total darkness manifest irreversible histological effects, such as degeneration of the optic disc, optic nerve, pyramidal cells in the striate area, and so on. Culturally disadvantaged children are obviously not reared in the dark. The experiments cited by Hunt are interesting but irrelevant to the problems discussed in my paper. Somewhat more relevant are Harlow's experiments (cited in my paper) on primates reared under severe sensory-motor deprivation but not the absence of light which results in optic-neural degeneration. Harlow's deprived monkeys were reared in isolation in small, lighted cages with uniform opaque walls and containing few manipulanda. Yet after prolonged periods of being raised in such an environment they showed no deficiencies in learning performance as compared with monkeys raised together in large, open cages permitting a variety of sensorimotor experience. Similar sensory deprivation and enrichment studies using rats, such as the work of Krech and Rosenzweig cited by Hunt, are clearly less relevant than the primate experiments in their implications for human behavior. It should be noted, however, that even in the case of rats, the greatest extremes of rat environment, from deprived to enriched (where the enrichment includes experience in mazes), that have been devised in the laboratory result in differences in maze learning ability only about one-fourth as large as those produced genetically by selective breeding for maze learning.

Hunt also attaches importance to experience in the development of sensorimotor integration, referring to experiments with rats climbing a guy-rope, which suggests that "each coordination, between vision-and-hand motion or between eye-function and ear-function, has its own neuro-electrical-chemical-anatomical equipment . . . When such equipment has emerged as the consequence of a given bit of functional accommodation or learning, it can readily be employed in other functioning and thereby becomes the basis for the transfer of training." But do such elemental components of sensorimotor accommodation and integration have any less chance to develop in a slum than in a penthouse? It seems far-fetched to me that, as Hunt suggests, these components of early sensorimotor development form the basis of Spearman's *g* or general intelligence factor. I cited evidence in my paper showing that, if anything, there is either a zero or a negative correlation between most indices of early behavioral development, such as the Bayley Infant Scales, and later IQ. Kagan (1966) has identified some components of early behavior which apparently show a more marked correlation with later intelligence than is generally found in the standard infant scales of development. Kagan reports that on certain laboratory tests of cognitive functioning lower-class children, as early as 8 to 12 months of age, show slower rates of information processing than middle-class children of the same ordinal position among their siblings. Kagan observes:

Lower-class children show less rapid habituation, less clear differentiation among visual stimuli, and, in a play situation, show a high threshold for satiation. The latter measure is obtained by placing the child in a standard playroom with a standard set of toys (quoits on a shaft, blocks, pail, mallet, peg board, toy lawn mower, and toy animals) and by noting the time involved in each activity. Some children play with the blocks for 10 seconds and then skip to the quoits or the lawn mower, playing only 10-20 seconds with each individual activity before shifting to another. A second group of children, called "high threshold for satiation infants" spends 1 or 2 minutes with an activity without interruption before changing. We do not believe the latter group of infants is taking more from the activity; rather it seems that they are taking longer to satiate on this action. It is important to note that the observation that lower-class infants show a high threshold for satiation contrasts sharply with the observation that 4-year-old lower-class children are distractible and hyperkinetic. We believe both descriptions. The paradox to be explained is why these lower-class children are pokey and lethargic and nondistractible at 12 months of age, yet display polar-opposite behaviors at 48 months of age (Kagan, 1966, pp. 105-106).

The other line of evidence appealed to by Hunt is on orphanage infants deprived of normal sensorimotor experience during the first one to two years of life, as in the well-known study by Skeels and Dye (1939). After such deprivation, these children have very retarded developmental quotients and their entire behavior is in marked contrast to that of children typically called disadvantaged. After placement in good environments, the children showed an average gain of about 30 IQ points, became average children, and grew up to be average adults (Skeels, 1966). This, too, is in contrast to typical disadvantaged children, who, rather than showing a tendency to catch up when placed in a presumably more culturally enriched environment—the school—begin gradually to fall behind in cognitive development. The typical characteristics of culturally disadvantaged children are a different set of phenomena from those resulting from early sensory deprivation. The contrast is further highlighted by studies of children who suffer severe verbal deprivation as a result of being born completely deaf. These children show a very marked retardation, usually amounting to one to two years, on tests of verbal intelligence. Unlike disadvantaged children, however, the deaf children, despite continuing deafness, gradually catch up in intellectual performance—it merely takes them longer to acquire information because of their severe sensory handicap. But once acquired, normal mental development continues. In one of the most careful studies of mental development in deaf children, the authors concluded that the deaf merely take *longer* to reach the same level of verbal-conceptual-thinking ability as normal persons. The authors state: ". . . the differences found between deaf and hearing adolescents were amenable to the effects of age and education and were no longer found between deaf and hearing adults. Dissociation between words and referents, verbalization adequacy, and level of verbalization were not different for deaf and hearing subjects. Our experiments, then, have shown few differences between deaf and hearing subjects. Those found were shown to fall along a normal developmental line and were amenable to the effects of increased age and experience, and education" (Kates, Kates, & Michael, 1962, pp. 31-32).

How much of Hunt's association of sensory deprivation with the culturally disadvantaged has affected psychologists' perceptions and descriptions of the environment of infants of mothers called culturally disadvantaged? Note Kagan's description of children he has studied in the lower-class white population: ". . . the lower class mothers spend less time in face to face mutual vocalization and smiling with their infants; they do not reward the child's maturational progress, and they do not enter into long periods of play with the child. Our theory of mental development suggests that specific absence of these experiences will retard mental growth and will lead to lower intelligence test scores." There is not unanimous agreement that the culturally disadvantaged have such impoverished interpersonal interactions in infancy as described by Kagan. The early environment of Negro infants, for example, is described in quite contrasting terms by a Negro writer, Kristin Hunter: "Ghetto babies must be the most thoroughly loved in the world; they are passed from loving arms to loving arms, cradled, cuddled, tickled, endlessly discussed and admired" (Hunter, 1969). This does not sound like sensory deprivation.

In emphasizing the environments of the extreme poor, Hunt remarks that "few if any of the studies of heritability have included the truly poor, so they have missed this portion of the variation in the circumstances of rearing." Heritability studies have included all social classes, but I agree that special attention should be given to including the very extremes of the existing environmental continuum. One might also expect, however, that sampling from an increased range of environments will simultaneously yield a correlated increase in genetic variation, thereby leaving the heritability of IQ approximately the same.

Hunt also seems to assume that anything that will accelerate any aspect of development is psychologically good and will have enhancing effects on later mental ability. This is sheer speculation without empirical support. Putting mobiles over a child's crib may very well bring about an earlier eye-blink response in infants, but what has this to do with the mental abilities measured by IQ tests and correlated with scholastic performance? There is just no evidence that these types of stimulation in early infancy, over and above what infants normally get, are in any way related to their intelligence at school age. In fact, there is some evidence, again from primate experiments, that attempting to develop abilities ahead of the normal maturation of cognitive processes may even be harmful. Harlow (1959), for example, found that very young monkeys have much greater difficulty than somewhat older monkeys in learning-set formation (i.e., "learning to learn") but that the younger monkeys can acquire learning sets by being given much more training than is needed by older monkeys. The younger monkeys, however, do not attain the same level of proficiency in these problems. The more important fact is that the younger monkeys cannot be trained to do as well as the older monkeys even when they finally reach the same age as the monkeys who trained at a later age. Harlow states: ". . . these data suggest that the capacity of the two younger groups to form discrimination learning sets may have been impaired by their early, intensive learning-set training, initiated before they possessed any effective learning-set capability." The more advanced cognitive structures awaiting later brain maturation apparently were never invoked in the earlier trained monkeys, whose performance remained permanently below that of monkeys trained at a later age. This observation would seem to be consistent with Elkind's conjecture that ". . . the longer we delay formal instruction, up to certain limits, the greater the period of plasticity and the higher the ultimate level of achievement."

Associative and Cognitive Abilities

My theory of two broad categories or clusters of mental abilities, labeled Level I and Level II because they seem to stand in some hierarchical relationship, is somewhat misinterpreted by Cronbach and Hunt. In factor analyses, a variety of tests of associative learning ability and memory (digit span, serial and paired-associate learning, free recall of uncategorized lists, etc.) tend to cluster together; these tests represent in varying degrees what I call Level I abilities. On the other hand, another class of tests, which are not highly correlated with Level I tests also cluster together: standard verbal and nonverbal IQ tests, tests involving abstract reasoning, symbol manipulation, free recall of conceptually categorized lists, etc. I call these abilities Level II.

Hunt lists a great variety of types of learning associated with various experimental techniques for the laboratory study of learning identified with Ebbinghaus, Pavlov, Thorndike, Hull, Skinner, and Piaget, and then says that my broad distinction between associative and cognitive learning is "but a conceptual drop in the bucket." This is to miss the point that Level I and Level II represent broad categories of abilities which do emerge in factor analyses, and many of the types of learning listed by Hunt can be represented in this two-dimensional factor space. The fact that one can fractionate these broad factors does not detract from their scientific usefulness in attempting to understand the structure of mental abilities. Nor is it meaningful to call this theory an "over-simplification" as does Cronbach. It *is* a simplification of a diversity of phenomena, to be sure, but an essential aim of science is to conceptually organize and simplify disparate and variegated phenomena. There is no doubt of the complexity inherent in my formulation. For example, few, if any, tests can be regarded as measuring purely Level I or Level II under all conditions. We already know that paired-associate learning tests can be either Level I or Level II, or any admixture of the two, depending upon a number of experimentally manipulable variables. For instance, if the subjects (college students) are forced to learn a list of paired-associates at a very fast rate of presentation, the test, when included in a factor analysis, is loaded almost entirely on the Level I factor. If the same paired-associates are presented at a much slower rate, the learning scores are then substantially loaded on the Level II factor. Also, certain instructional techniques may change what are usually perceived as rote-learned tasks into conceptually mediated learning. Cronbach should be assured that I recognize a *continuum* of the susceptibility of various tasks to manipulation with respect to their Level I-Level II loadings. Some tasks are relatively easy to manipulate in this respect—for example, paired-associate learning and probably free recall of clusterable lists. Other tasks are much more difficult to manipulate through instruction, for example, the ability of children under 6 or 7 to copy the figure of a diamond, or to conserve volume in the Piagetian paradigm.

All this does not mean, however, that stable individual differences in Level I and Level II abilities do not exist or are trivial. Cronbach points out that spatial ability, which is highly heritable, can be improved through training. I hope he does not believe that this implies that the training will wipe out, or even decrease, individual differences in spatial ability or will lower its heritability within the group that received the training. There is good reason to believe that just the opposite would occur. I have found in some of my own research, for example, that prolonged practice (by college students) on digit span tests significantly *increases* the amount of reliable variance due to individual differences. All subjects improve with practice, but reliable individual differences become accentuated at the asymptote of improvement. Cronbach knows that when we talk about the heritability of an ability, we are not referring to the absolute level of performance that can be attained, but to individual differences in performance and the proportion of their variance attributable to genetic differences.

The Hope of the Instruction × Individual Differences Interaction

Hunt and especially Cronbach share the same hope I expressed in my paper (and on numerous other occasions) that the improvement of scholastic achievement and the minimization of individual and group differences in performance may be brought about by making use of the idea of a subjects × instruction interaction. In the simplest terms this means that if Jim and Bill are taught in the same way, they will differ more in how fast and how much they learn than they would if each one were taught by a different method which is especially suited to each child's individual pattern of abilities. Bereiter is clearly much less optimistic than the rest of us about the practical possibilities implied by the instructional interaction notion. His cogent remarks have indeed had a somewhat sobering effect on my own thinking on this topic and I have gone back to the literature to see how much hard evidence I could find to bolster my hope that this interaction notion of more individualized instruction holds the promise of solving our major educational problems. To my dismay, but in all fairness to Bereiter, I must admit that I can find very little evidence of pupil × type of instruction interaction in the realm of learning school subjects or for complex learning in general. Most of the evidence for such pupil × instruction interactions has been reviewed by Cronbach (1967) in a paper which is a "must" in this field. I believe that research based on a more fine-grained approach to the analysis and manipulation of instruction will be necessary before we can properly assess the educational potential of the pupil × instruction interaction. We do know that quite clear-cut interactions have been shown in laboratory experiments on simple learning tasks in which the tasks and methods themselves impose great constraints on what the subject can do in the learning situation. Then we can find significant interactions between learners and experimental variables (Jensen, 1967). When tasks are complex, involving a variety of abilities, as in school learning, and when there are few constraints on how subjects can learn, pupil × instruction interactions either fail to appear or are undetectable. At this point, indeed, I can only say it is my conjecture, my hope, that the Level I-Level II distinction may interact with instructional techniques to decrease the spread between disadvantaged and advantaged children in their mastery of the basic scholastic skills. I hope a variety of research will be directed to testing this hypothesis.

Cronbach solves no problem by saying "Capability is not at issue when a child does not call upon an ability he possesses." What about the ability to call up relevant subabilities and past learning when confronted with a new problem? This ability to transfer learning from one type of problem to another is the essence of intelligence; it is a Level II process. Why does the 5-year-old fail to copy a diamond despite his ability to draw straight lines? Why does a child who has learned to add, subtract, multiply, and divide often fail in arithmetic "thought problems" which call upon the applications of these subabilities? It is the appropriate calling up, integration, and transfer of various subskills that constitute what we mean by intellectual capability. I can play chess; I know all the moves. But why can't I play like Alekhine or Capablanca? Is it simply because I do not call upon an ability that I possess? I doubt it.

Bereiter is correct, I believe, in his argument that complex intellectual tools act as amplifiers rather than equalizers of basic differences in problem-solving ability. Cronbach's argument that the invention of the computer has increased man's mathematical capacity has as much to do with individual differences in mathematical ability as the invention of the automobile has to do with individual differences in running ability.

Genetic Social-Class Differences in Intelligence

Because of differences between child-rearing practices of the middle-class and those of people of poverty, Hunt doubts that socioeconomic status (SES) differences in intelligence have any genetic component. If Hunt's supposition were true that there is no genetic component to social class intelligence differences, it would have to mean that all the factors involved in social mobility, educational attainments, and the selection of persons into various occupations have managed scrupulously to screen out all variance associated with genetic factors among individuals in various occupational strata. The possibility that the selection processes lead to there being only environmental variance in intelligence among various socioeconomic groups and occupations—a result that could probably not be accomplished even by making an explicit effort toward this goal—is so unlikely that the argument amounts to a *reductio ad absurdum*. If individual differences in intelligence are due largely to genetic factors, then it is virtually impossible that average intelligence differences between social classes (based on educational and occupational criteria) do not include a genetic component.

The argument is as follows: Twin studies and other methods for estimating the heritability of intelligence have yielded heritability values for the most part in the range from .70 to .90, with a mean value of about .80. Heritability (H) indicates the proportion of variance in a metric characteristic, such as height or intelligence, that is attributable to genetic factors. (Since the heritability estimate is derived from studies in European and North American Caucasian populations, the present genetic analysis of SES differences cannot be generalized across racial groups.)

$1 - H = E$, the proportion of variance due to non-genetic or environmental factors, which of course include prenatal as well as postnatal influences. The correlation between phenotypes (the measurable characteristic) and genotypes (the genetic basis of the phenotype) is the square root of the heritability, i.e., $\sqrt{H}$. An average estimate of $\sqrt{H}$ for intelligence is .90, which is the average correlation between genotype and phenotype. An estimate of the average correlation between occupational status and IQ is .50. What Hunt is saying, essentially, is that the correlation between IQ and occupation (or SES) is due entirely to the environmental component of IQ variance. In other words, this hypothesis requires that the correlation between genotypes and SES be zero. So we have correlations between three sets of variables: (a) between phenotype and genotype, $r_{pg} = .90$; (b) between phenotype and status, $r_{ps} = .50$; and (c) the hypothesized correlation between genotype and status, $r_{gs} = 0$. The first two correlations (r_{pg} and r_{ps}) are determined empirically and are represented here by average values reported in the literature. The third correlation (r_{gs}) is hypothsized to be zero by those who believe genetic factors play a part in *individual* differences but not in SES *group* differences. The question then becomes: is this set of correlations possible? The first two correlations we know are possible because they are empirically obtained values. The correlation seriously in question is the hypothesized $r_{gs} = 0$. We know that mathematically the true correlations among a set of variables, 1, 2, 3, must meet the following requirement:

$$r^2_{12} + r^2_{13} + r^2_{23} - 2r_{12}r_{13}r_{23} < 1$$

The fact is that when the values of $r_{pg} = .90$, $r_{ps} = .50$ and $r_{gs} = 0$ are inserted into the above formula, it yields a value greater than 1.00. This means that r_{gs} must in fact be greater than zero.

Another way of regarding this problem is as follows: If only the E (environmental) component determined IQ differences between status groups, then the H component of IQ's would be regarded as random variation with respect to status. Thus, in correlating IQ with status, the IQ test in effect would be like a test with a reliability of $1 - H = 1 - .80 = .20$. That is to say, only the E component (.20) of the total variance is not random with respect to indices of SES. Therefore the theoretical maximum correlation that IQ could have with SES would be close to $\sqrt{.20} = .45$. This value is slightly below but very close to the average value of obtained correlations between IQ and SES. So if we admit no genetic component in SES differences, we are logically forced to conclude that persons have been fitted to their socioeconomic status (meaning largely educational attainments and occupational status) almost *perfectly* in terms of their environmental advantages and disadvantages. In other words, it would have to be concluded that persons' innate abilities, talents, and proclivities play no part in educational and occupational selection and placement. This seems a most untenable conclusion. The only way one can logically reject the alternative conclusion, that there are average genetic intelligence differences among SES groups, is to reject the evidence on the heritability of individual differences in intelligence. But the evidence for a substantial genetic component in intellectual differences is among the most consistent and firmly established research findings in the fields of psychology and genetics.

Social and Educational Policy and the Heritability of Individual Differences

Cronbach states it is regrettable that I do not spell out the policies that should follow from my formulations and conclusions. This is, of course, another job. I am not a social or educational philosopher and I am sure that neither I nor anyone else at present has thought through all the policy implications of my article. I do believe that educational policy decisions should be based on evidence and the results of continuing research—and not just the evidence which is comfortable to some particular ideological position, but *all* relevant evidence. I submit that the research on the inheritance of mental abilities *is* relevant to understanding educational problems and formulating educational policies. For one thing, it means that we take individual differences more seriously than regarding them as superficial, easily-changed manifestations of environmental differences. And it means we look more critically and carefully at environmental variables that contribute most to differences in mental development, as I suggested that prenatal and nutritional factors had not been given due consideration. Also, it means we expend more research effort on exploring and mapping a wider range of abilities than those measured by IQ tests, on discovering the particular learning strengths of each child, and on devising methods that will more fully utilize these strengths to help all children to benefit more from their schooling. To refrain from discussing some of the relevant factors that should be considered in formulating policy simply because the details of such policy cannot yet be spelled out is, in my opinion, practically equivalent to saying: "Don't ask any questions unless you already know all the answers."

Brazziel's letter seems to be saying in part that my paper should not have been published in the first place. I would plead for more faith in the wisdom of the First Amendment. To refrain from publishing discussions of research on socially important issues because possibly there will be some readers with whose interpretation or use of the material we may disagree is, in effect, to give those persons the power of censorship over the publication of our own questions, findings, and interpretations. It is only when all the available facts, issues, and questions can be openly examined and discussed by everyone that we can put any stock in the maxim that "the truth will out." I resent Brazziel's statement that I expound a theory of white supremacy, but I suppose it must be evaluated in the context of his overall reaction to my article. On this point, however, it might be of interest to some to note that on the basis of the evidence I have been able to review so far, if I were asked to hypothesize about race differences in what we call g or abstract reasoning ability, I would be inclined to rate Caucasians on the whole somewhat below Orientals, at least those in the United States. A case can be made for this conjecture on the basis of existing evidence, but this is not the appropriate place for it.

Reducing the Uncertainties

One disappointment with the discussions of my paper is the fact that attitudes of "let's not talk about genetics," or "it's too complicated," or "we can't find out the answers anyway," and so on, have prevailed over the attitude of inquiry and the application of intellectual ingenuity in trying to reduce our heredity-environment uncertainty. If there are weaknesses in the methods and the evidence I have presented, and of course there inevitably are at this stage, we would do well to note them as a basis for seeking more refined research methods and more and better data, rather than as a basis for minimizing the scientific and social importance of these questions, or sweeping them under the rug.

Brazziel is quite correct in noting, for example, that the Negro population of the United States, like the white, is very far from being genetically or racially homogeneous. In fact, it is doubtful that any babies of pure African descent are being born in the United States today, unless they are born to African exchange students. But Africans, too, are genetically heterogeneous. A number of studies based on the differential frequencies of various blood groups in African and Caucasian populations have shown that, on the average, persons socially classified as American Negroes now have an admixture of 20 to 30 per cent Caucasian genes (Reed, 1969). The percentage of Caucasian admixture varies greatly in various regions of the country, going from an average of below 10% in some Southern states to above 25% in some Northern states. These figures can be estimated with considerable precision in large population samples, depending on the number of different blood groups and other genetic polymorphisms one is able to take into account. With these methods individuals, too, can be categorized by proportions of Negro-Caucasian admixture on a probabalistic basis. Possibly these same genetical techniques could provide a basis for more refined and accurate tests of hypotheses concerning racial differences in ability patterns. Since skin color is but poorly correlated with the percentage of Caucasian admixture, and because it may have social-environmental consequences, it could be statistically controlled in studies of the correlation between Negro-Caucasian admixture and measures of psychological characteristics. Environmental differences would not be an obstacle, since there is a wide range of racial admixtures in any large sample from highly similar environments. In fact, where there are half-siblings, intra-family comparisons might be possible, thereby controlling a host of environmental family-background factors. Other quite different approaches are possible, or a number of methods used in combination. The finding that electroencephalographic visually-evoked potentials are related to IQ means that intelligence might be measured on a physiological level, and such a measure would come closer than anything we now have to a true culture-free test. Studies of foster children of one race or social class adopted by parents of another is one more avenue. Such are only a few of the possible suggestions. Geneticists should be able to evaluate these and come up with better ideas. Collaborative research by geneticists and behavioral scientists could surely advance our scientific knowledge of racial and social class differences. To argue to the contrary, it seems to me, is to claim the impotence of a scientific approach and of human ingenuity, an attitude which is clearly contradicted by our great advances in other fields of inquiry. If the heredity-environment uncertainty is unresolvable in the sense that, say, perpetual motion is impossible, we should at least not be satisfied until we have discovered precisely the laws of nature which make it so.

It is already apparent that my article "How Much Can We Boost IQ and Scholastic Achievement?" has been eminently successful in widely provoking serious thought and discussion among leaders in genetics, psychology, and education concerning important fundamental issues and their implications for education. I expect now that this will stimulate further relevant research as well as efforts to apply the knowledge gained thereby to educationally and socially beneficial purposes. The whole society will benefit most if scientists and educators treat these problems in the spirit of scientific inquiry rather than as a battle field upon which one or another preordained ideology may seemingly triumph.

References

Bloom, B. S. *Stability and change in human characteristics*. New York: Wiley, 1964.

Carter, C. O. Differential fertility by intelligence. In J. E. Meade & A. S. Parkes (Eds.), *Genetic and environmental factors in human ability.* New York: Plenum Press, 1966. Pp. 185-200.

Cronbach, L. J. How can instruction be adapted to individual differences? In Gagné, R. M. (Ed.), *Learning and individual differences.* Columbus, Ohio: Merrill, 1967. Pp. 23-39.

De Lemos, M. Murray. The development of the concept of conservation in Australian aboriginal children. Unpublished Ph.D. dissertation. University of Western Australia, Nov., 1966.

Fuller, J. L., and Thompson, W. R. *Behavior genetics.* New York: Wiley, 1960.

Gottesman, I. I. Biogenetics of race and class. In M. Deutsch, I. Katz, & A. R. Jensen (Eds.), *Social class, race, and psychological development.* New York: Holt, Rinehart & Winston, 1968. Pp. 11-51.

Harlow, H. F. The development of learning in the Rhesus monkey. *Amer. Sci.,* 1959, **47,** 459-479.

Hunter, Kristin. Pray for Barbara's baby. *Philadelphia Magazine,* Aug., 1968. (Reprinted in *Reader's Digest,* January, 1969).

Jensen, A. R. Varieties of individual differences in learning. In Gagné, R. M. (Ed.), *Learning and individual differences.* Columbus, Ohio: Merrill, 1967. Pp. 117-135.

Jensen, A. R. Social class, race, and genetics: Implications for education. *Amer. Educ. Res. J.,* 1968, **5,** 1-42. (a)

Jensen, A. R. Another look at culture-fair tests. In *Western Regional Conference on Testing Problems, Proceedings for 1968,* "Measurement for Educational Planning." Berkeley, Calif.: Educational Testing Service, Western Office, 1968. Pp. 50-104. (b)

Kagan, J. A developmental approach to conceptual growth. In H. J. Klausmeier & C. W. Harris (Eds.), *Analyses of concept learning.* New York: Academic Press, 1966. Pp. 95-115.

Kates, Solis L., Kates, W. W., & Michael, J. Cognitive processes in deaf and hearing adolescents and adults. *Psychol. Monogr.,* 1962, **76,** Whole No. 551.

Kohlberg, L. Early education: A cognitive-developmental view. *Child Developm.,* 1968, **39,** 1013-1062.

Minor, J. B. *Intelligence in the United States.* New York: Springer, 1957.

Reed, T. E. Caucasian genes in American Negroes. Unpublished manuscript, March, 1969.

Scott, J. P., & Fuller, J. L. *Genetics and the social behavior of the dog.* Chicago: Univer. of Chicago Press, 1965.

Shields, J. *Monozygotic twins brought up apart and brought up together.* London: Oxford Univer. Press, 1962.

Skeels, H. M. Adult status of children with contrasting early life experiences: A follow-up study. *Child Developm. Monogr.,* 1966, **31,** No. 3, Serial No. 105.

Skeels, H. M., & Dye, H. B. A study of the effects of differential stimulation on mentally retarded children. *Proc. Addr. Amer. Ass. Ment. Defic., 1939,* **44,** 114-136.

Tanner, J. M. Earlier maturation in man. *Sci. Amer.,* 1968, **218,** 21-28.

Tanner, J. M. The trend towards earlier physical maturation. In J. E. Meade & A. S. Parkes (Eds.), *Biological aspects of social problems.* New York: Plenum Press, 1965. Pp. 40-66.

Tuddenham, R. D. Soldier intelligence in World Wars I and II. *Amer. Psychol.,* 1948, **3,** 54-56.

Tuddenham, R. D. Psychometricizing Piaget's methode clinique. Paper read at Amer. Educ. Res. Ass., Chicago, February, 1968.

Vernon, P. E. Environmental handicaps and intellectual development: Part II and Part III. *Brit. J. educ. Psychol.,* 1965, **35,** 1-22.

Notes on Contributors

CARL BEREITER, specialist in cognitive development, is Professor of Applied Psychology at the Ontario Institute for Studies in Education. Co-author, with Siegfried Englemann, of *Teaching Disadvantaged Children in the Preschool* (1966), his most recent book is *Must We Educate?* (1973).

WILLIAM F. BRAZZIEL is Professor of Higher Education at the University of Connecticut. A psychologist, he has served as Director of General Studies at Norfolk State College. He has also held teaching and administrative posts at Southern University at Baton Rouge, Louisiana, and Central State College in Wilberforce, Ohio. Author of many articles, he has done research on early education and training the hard-core unemployed. Most recently, he is author of *Quality Education for All Americans* (1974).

LEE J. CRONBACH, Vida Jacks Professor of Education at Stanford University, is a specialist in educational psychology and evaluation. Editor, with P. J. Drenth, of *Mental Tests and Cultural Adaptation* (1972), he is also one of the many authors of *Dependability of Behavioral Measurements* (1972).

JAMES F. CROW is Professor of Genetics at the University of Wisconsin, Madison, where he was previously Acting Dean of the School of Medicine. Currently concerned with population genetics, he is the author of *Genetics Notes* (1966), and co-author of *An Introduction to Population Genetics Theory* (1970).

DAVID ELKIND is Professor of Psychology, Psychiatry and Education at the University of Rochester where he is Director of the Graduate Training Program in Developmental Psychology. He is primarily concerned with cognitive and perceptual development in children and adolescents, parent-child relations, and delinquency. He is co-author, with Irving B. Weiner, of *Child Development: A Core Approach* (1972), *Children and Adolescence: Interpretative Essays on Piaget* (2nd ed., 1974), and *Sympathetic Understanding of the Child: Six to Sixteen* (2nd ed., 1974).

J. McV. HUNT is Professor Emeritus of Psychology and Education at the University of Illinois, Urbana. He was Chairman of President Johnson's Task Force on Early Childhood and authored the report, *A Bill of Rights for Children* (1967). He has received a Research Career Award from the National Institute of Mental Health for a study of the roots of intelligence and motivation in early experience. He is author of *Intelligence and Experience* (1961), *The Challenge of Incompetence and Poverty* (1969), and co-author, with Ina C. Uzgiris, of the forthcoming *Assessment in Infancy: Ordinal Scales of Psychological Development* (1975).

ARTHUR R. JENSEN, Professor of Educational Psychology at the University of California at Berkeley, is the author of numerous articles on individual differences in human learning, serial learning, and cultural, developmental and genetic determinants of intelligence and learning ability. His most recent publications are *Genetics and Education* (1973), *Educability and Group Differences* (1973), and *Educational Differences* (1973).

JEROME KAGAN, Professor of Developmental Psychology at Harvard University, is a specialist in early childhood cognitive development including cross-cultural comparisons. Among his publications are, with H. A. Moss, *Birth to Maturity* (1962), *Change and Continuity in Infancy* (1971), *Understanding Children* (1971), and with P. H. Mussen and J. J. Conger, *Child Development and Personality* (4th ed., 1974).